マスターしておきたい
技術英語の基本
― 決 定 版 ―

Richard Cowell
佘 錦華
共 著

コロナ社

Preface

With the increasing interchange among nations and the advance of businesses into foreign countries, the ability to write technical English has become indispensable to researchers and engineers. To meet this need, many books have been published on how to write technical English. Most of them take the form of showing concrete examples of incorrect sentences and then the correct English. In contrast, we think it is also very important to consider words and expressions that readers may often misuse, and not only help them to understand the reason for the mistake and the correct usage, but at the same time provide guidance in writing more natural English that gets away from the Japanese way of thinking. However, it is almost impossible to find any books written from this point of view. In light of this situation, the authors decided to put out a book on writing technical English from this new viewpoint.

This book provides guidelines on the usage of words and expressions that are often misused by Japanese researchers, and some basic tips on style and punctuation. Since it contains only guidelines, it is not an exhaustive study. That is, if you follow the guidelines, your sentences will be correct; but in articles written by native speakers of English, you may come across sentences that do not follow the suggestions in this book. Furthermore, unlike other books, this one contains exercises after many of the topics to enable readers to check their level of understanding. To use the book, readers do not need a detailed knowledge of English grammar; a familiarity with basic grammar will suffice.

Two key aspects of any word or expression are **meaning** and **usage**. It seems that there is a common tendency for people to look up the **meaning** of a word in an English-Japanese dictionary, and then to feel that they can use the word freely in an English sentence. This is the cause of many mistakes, because knowing the meaning of a word just gives you the ability to **read** it. In other words, when you try to write a sentence with it, you naturally follow the Japanese usage. In order to **write** correctly, you must also know the **usage**. This book is filled with guidance on correct usage.

The topics covered in this book are the result of over 20 years of experience correcting technical papers written in English by Japanese

Preface

researchers, and the realization that the same mistakes keep appearing again and again. The book is loosely organized with the most common problems at the beginning and less frequently occurring problems at the end.

Almost all of the example sentences come from the fields of electronics and information technology. However, readers not familiar with these fields can still obtain great benefit from this book by substituting simple words for the very technical terms and focusing on usage patterns. For example, the following sentences from the topic "apply" in Section 1 illustrate a typical mistake and how to correct it:

- ✗ Many attempts have been made to *apply* the microphase-separated domains of block copolymer *as* a dry etching mask.
- ○ Many attempts have been made to **use** the microphase-separated domains of block copolymer **as** a dry etching mask.

Extracting the pattern from the sentence, we obtain the following.

- ✗ ...to *apply* (a substance) *as* (something)
- ○ ...to **use** (a substance) **as** (something)

In fact, all readers should focus on usage patterns so that they can apply them to their own writing.

The authors would be very pleased if the readers found this book at all useful in their writing of technical English.

Finally, the authors are deeply grateful to the Corona Publishing Co. Ltd for undertaking the publication of this book. We would also like to express heartfelt appreciation to Mr. Shiro Morita for his careful checking of the manuscript and for his valuable advice. Furthermore, we are grateful to Hiroyuki Kobayashi (Faculty of Engineering, Osaka Institute of Technology), Ying Li (Atomic Frequency Standards Group, National Institute of Information and Communications Technology), and Yi Ren (Department of Chemistry and Biotechnology, School of Engineering, the University of Tokyo) for reading the manuscript. And we would like to thank Ying Li for her advice on the design of the text.

April, 2006

Richard Cowell & Jinhua She

まえがき

　国際交流の増加と企業の海外進出に伴い，技術英語の作文能力は，研究者・技術者にとって欠かせないものとなっている。そのニーズに応えるために，技術英語の書き方についての成書が多数出版されている。その多くは，間違い文例を具体的に示し，それを正しい英語に直していく形で構成されている。これとは別に読者が使い方を誤りやすい単語や表現をとり上げて，それぞれの誤りの理由と，正しい使い方を理解させると同時に，日本語的な発想から離れ，より自然な英文を書くことへの，道案内をすることも大切と思われる。しかしながら，このような見地から書かれた書物は，ほとんど見当たらない。著者らはこのような現状に照らして，この新しい見地から，技術英語の書き方に関する本書を著わすことにした。

　本書では，日本人が間違えやすい，単語と表現の使い方への指針と，スタイルおよび句読法に関する基本的なヒントをまとめている。ガイドライン的に構成されているため，あらゆる事柄をすべて網羅しているわけではない。すなわち，本書のアドバイスに従うと，読者は正しい文書を書くことができるが，英語を母語とする人の作成した文章の中には，本書の指針に従わないものもあるかもしれない。また，従来の成書と違い，理解度が確認できるように，重要な項目の後に，練習問題も用意した。本書を活用するうえで，英文法の詳しい知識は必要ではなく，基本的な文法を習得していれば十分である。

　単語および表現は**意味**と**使用法**という二つの側面を持っているが，多くの人々は英和辞書で単語の意味を調べただけで，それが自由に使えると思っているようである。しかし，単語の**意味**を知ることは，ただそれに関する**読解力**を身につけただけに過ぎないため，文章を書こうとすると多くの間違いを犯すことになる。すなわち，その単語を用いて文章を作成することになると，つい日本語式の使い方で英文を書いてしまう。正しい文章を**書く**ためには，その単語の本来の**使用法**をマスターする必要がある。本書には，単語の正しい使い方の手引きが数多く述べられている。

　著者の一人は，20年以上にわたり英文の技術論文を添削している間に，まったく同じ間違いが繰り返し現れることに気づいた。その経験に基づいて本

書のトピックを選び,一番間違いやすい単語・表現から順に,巻頭から巻末にかけて並べた.

本書中に使用される例文は,おもに情報・電子分野のものであるが,これらの分野に馴染みのない読者でも,専門用語を単純な単語に置き換え,**使用パターン**に注目すれば,本書から多くのことが得られると確信している.例えば,**Section 1**の「**apply**」に,以下のような典型的な間違いと,その添削を示している.

- × *Many attempts have been made to <u>apply</u> the microphase-separated domains of block copolymer <u>as</u> a dry-etching mask.*
- ○ Many attempts have been made to **use** the microphase-separated domains of block copolymer **as** a dry-etching mask.

上の文のパターンをとり出してみると

- × *...to <u>apply</u> (a substance) <u>as</u> (something)*
- ○ ...to **use** (a substance) **as** (something)

このように,すべての読者は,自分の作文に応用できる**使用パターン**に注目することを勧める.

本書が,少しでも読者の技術英語作文に役立つことができれば幸いである.

最後に,本書の出版を引き受けてくださったコロナ社に心よりお礼を申し上げます.また,原稿を入念にチェックし貴重な助言をしていただいた森田司郎氏に深く御礼を申し上げます.草稿を読んでいただいた大阪工業大学工学部の小林裕之氏,独立行政法人情報通信研究機構電子波計測部門の李瑛氏,東京大学工学部化学生命工学科の任宜氏に謝意を表します.なお,李瑛氏からテキストのデザインに関するアドバイスもいただいたことを感謝するしだいです.

2006年4月

著者らしるす

Preface to the definitive edition

Although the English proficiency of Japanese people continues to rise, writing technical English still poses special problems. The style, grammar, punctuation, and vocabulary must be good enough for publication in an international journal. Good writing can only be learned by actually writing, just as speaking a foreign language can only be learned by speaking. Reading a lot of well-written English can help a great deal by giving one a "feel" for the language. But inevitably, when a non-native speaker begins to write English, there will be gaps in his knowledge of the language. Most people fill in those gaps by relying on their native language and directly translating words, phrases, and style. This is the origin of a great number of the mistakes made by Japanese people who write manuscripts in English. Since these mistakes "echo" their own native language, they sound somewhat natural and are difficult to detect. Moreover, most people tend to make the same kinds of mistakes because the origin is the same. This book is a collection of some of those mistakes, accumulated over years of correcting technical English.

This edition of the book follows the same basic pattern as the first edition, with the most frequently made mistakes tending to be near the beginning, but there are some differences.

(a) New topics have been added, and some new explanations have been written for old topics (See especially the topic "respectively" in Section 4).

(b) In the first edition, examples usually consisted of pairs of incorrect and correct sentences. In this edition, the pairs have often been condensed into a single sentence, with extensive use being made of the strikethrough to indicate an incorrect expression. This saves a great deal of space.

(c) An attempt has been made to increase the density of information per page by filling in the empty space on each page with small items.

(d) The examples and practice sentences have been simplified a little to make them easier to understand.
(e) A new section has been added that contains hints for making an oral presentation. It includes a couple of pages on how to chair a session of an international conference.
(f) A full range of colors is used.

    The authors wish to thank the readers who found the first edition to be useful; and we hope that the definitive edition will be even more useful and that even more people will find some benefit in reading it.

    Finally, we are deeply grateful to the Corona Publishing Co. Ltd for encouraging the writing of the definitive edition and for waiting so patiently for it to be finished. We would also like to express heartfelt appreciation to Mr. Shiro Morita, Mr. Ichiro Fujii, Dr. Kohji Makino of the Faculty of Engineering, University of Yamanashi, Dr. Daisuke Chugo of the School of Science and Technology, Kansei Gakuin University, and Kou Miyamoto, a graduate student at the Tokyo Institute of Technology for their careful checking of the manuscript and for their valuable advice.

July 2015

Richard Cowell & Jinhua She

決定版のまえがき

　日本人の英語力は向上しているが，技術英語を書くことにはまだ課題が多い。文のスタイル，文法，句読点と語彙は，国際ジャーナルに受理されるためには十分なレベルでなければならない。外国語を話すことを学ぶ唯一の方法は，実際に話すことであるのと同じように，よい文書を作成することを学ぶ唯一の方法は実際に書くことである。それにしても，英語の「感触」に慣れるためには，たくさんの，よく書かれた英語論文を読むのが有益である。英語を母国語としない人が英語を書こうとすると，正しい英語の知識が不足するのはやむを得ない。多くの日本人はその知識不足を補うため，意味が似ていて，頭に浮かぶ日本語を英語に直訳し，それを使って論文を書こうとする。英語原稿を作る日本人が書く多くの間違いの起源がここにある。この種の間違いは，もともと自国語が起源なので，日本人にとっては自然に見えて，かえって誤りであることを見つけにくい。さらに，起源が同じであるので，多くの日本人は同じ種類の間違いをする傾向がある。この本の初版，決定版ともに，長年にわたる技術英語の添削によって蓄積された間違いのコレクションである。

　初版では頻繁に起こる間違いを，本の最初の方のページにおいた。決定版も同様な方針であるが，多少の変更が加えられている。

(a) 新しいトピックと説明が追加された（第4節で「respectively」の話題はその例である）。
(b) 初版において，文例としては，通常誤った文と正しい文のペアが並んでいた。決定版では，そのペアはしばしば一つの文にまとめられて，不適切な表現を示すために，取り消し線を使用した。これによって，必要なスペースを大いに減らすことができた。
(c) ページにある空きスペースを，小さなトピックに割り当てることによって，1ページあたりの情報密度を増加させた。
(d) 例文と練習のための文は，理解しやすくするために，少々簡単にされた。

(e) 口頭発表に役立つヒントを集めた新しいセクションを追加した。国際会議のセッションの司会を務める方法も説明された。

(f) 初版に比べてより多くの色が使われた。

著者らは，これは役に立つと喜んだ初版の読者に感謝します。私たちは，この決定版で新たな読者のみなさんを歓迎するとともに，旧版からの読者にも一層役に立つであろうことを願っています。

最後に，決定版の出版を励まし辛抱強く待ってくださったコロナ社に心よりお礼を申し上げる次第である。また，原稿に目を通し貴重な助言をくださった森田司郎氏，藤井一郎氏，山梨大学工学部の牧野浩二氏，関西学院大学理工学部の中後大輔氏と東京工業大学大学院の宮本皓氏に深謝したい。

2015年10月

著者らしるす

Contents

Section 1 — 1
- in this work vs. in this paper — 2
- novel — 3
- ave vs. avg — 3
- realize — 4
- confirm — 7
- that vs. which — 9
- Adjective Clause: Short Form — 11
- Changes & Differences — 14
- first vs. at first — 17
- operating principle — 18
- evaluate vs. estimate — 19
- Units — 21
- enable — 22
- Punctuation: Space — 23
- Style: Dynamic Verbs 1 — 25
- Prepositions 1 — 26

Section 2 — 27
- propose — 28
- Lists — 30
- Specifying Values — 35
- depend on — 35
- contain vs. include — 36
- on the contrary — 39
- adopt — 40
- cannot can not can't — 41
- in case of fire — 42
- Connecting Nouns — 44
- apply — 46
- Punctuation: Hyphen (-) — 48
- Style: Dynamic Verbs 2 — 50
- Prepositions 2 — 51

Section 3 — 52
- compared to vs. than — 53
- damage vs. damages — 54
- for -ing — 55

x Contents

is expected	57
approach	58
can could	59
consist of	61
as a result	62
proportion(al)	63
is thought	64
each	65
prepare	66
becomes vs. is	67
Punctuation: Colon (:)	69
Style: Unnecessary Repetition	70
Prepositions 3	71

Section 4　　　　　　　　　　　　　　　　　72

remarkable	73
control	74
tolerance	75
respectively	76
common vs. popular	78
recently	79
simplified	80
introduce	81
enough	82
Adjective Formation (-ing)	83
Adjective Formation (-ed)	84
compensate：他動詞 vs. 自動詞	86
conventional	86
Punctuation: Comma 1	87
Style: Unnecessary Words 1	89
Prepositions 4	90

Section 5　　　　　　　　　　　　　　　　　91

effective	92
has been used vs. is used	94
number	96
by vs. with	98
the both, the each, the another...	99
Keep Related Words Together	100
multi-	102
fixed	103
coincide	105

traditional	105
correspond	106
Punctuation: Comma 2	108
Style: Unnecessary Words 2	110
Prepositions 5	111

Section 6 — 112

know vs. find out	113
approach to key to	114
then	115
a/an vs. one of (the)	116
most vs. most of (the)	117
none, one, some, most, all	118
Meaningless —ing	120
issue vs. problem	121
obvious	123
so-called	123
optics is vs. optics are	124
therefore vs. so	125
problem with/of	126
Punctuation: Semicolon (;)	127
Style: (Fig. 3)	129
Prepositions 6	130

Section 7 — 131

measured vs. measurement	132
contribute to	132
Bad Passives	133
because vs. since	134
composition vs. content	135
another vs. the other	136
maintain vs. remain	137
improve	138
saturate	139
not A or B	141
be consistent with	142
recover vs. restore	143
monotonous vs. monotonic	144
Punctuation: Parentheses	145
Style: Larger For A Than For B	146
Prepositions 7	147

Section 8 — 148

fluctuations vs. variation	149
however, then, therefore, thus	150

with increasing frequency … 151
almost … 152
whose … 153
performance vs. performances … 154
flow … 155
against … 156
complete(ly) vs. perfect(ly) … 157
summarize … 159
reach … 159
small AND red? … 160
commercialized, specialized, standardized … 161
compare between … 162
express … 163
XXXable … 164
Change vs. Comparison … 166
Punctuation: Slash … 167
Punctuation: Capitals … 168
Punctuation: Dash … 168
Prepositions 8 … 169

Section 9: 口頭発表のヒント … 170

形式のレベル … 171
初めに … 171
アウトライン … 172
次のトピックへの移行 … 172
聴衆を見ること … 173
I vs. We … 173
略語の導入 … 173
スライド … 174
レーザーポインター … 176
終わり … 176
暗 記 … 176
プレゼンテーションを短くする方法 … 177
質疑応答の時間 … 178
練 習 … 179
セッションの司会を務める方法 … 181

References … 184
Answers … 185
Index … 202

Section 1

in this work vs. in this paper

novel

ave vs. avg

realize

confirm

that vs. which

Adjective Clause: Short Form

Changes & Differences

first vs. at first

operating principle

evaluate vs. estimate

Units

enable

―――――――――――

Punctuation: Space

Style: Dynamic Verbs 1

Prepositions 1

in this work vs. in this paper

Research (研究)

「Work」と「research」などの言葉は普通研究室で行われる**研究活動**を意味する。その活動は通常，**過去時制**で記述される。

研究を説明するのに使う動詞
analyze, calculate, create, develop, devise, design, determine establish, estimate, evaluate, fabricate, invent, investigate, make, measure, model, test, use, etc.

Reporting (報告)

「Paper」，「report」，「article」，「letter」，「presentation」などの言葉は研究に関する**報告**を意味する。その内容は通常，**現在時制**で記述される。

報告を説明するのに使う動詞
describe, discuss, explain, present, propose, report (on), review, etc.

NOTE: 「**Study**」は研究自体または研究に関する報告書という二つの意味がある。同じ段落中に両方の意味を使ってはいけない。

概要または要約を書くときに，研究または報告のどちらか一つの立場をはっきりさせ，同じ意味を通して使用すること。

Good Examples

In this **work**, the rating of perceived exertion **was used to determine** the maximum pedal load for an electric cart for the elderly. A dynamic cart control system that guarantees robust stability **was designed**. In addition, a stability condition based on the concept of dynamic parallel distributed compensation **was established**. Finally, experiments **were carried out** to demonstrate the validity of the method.

This **paper** first **discusses** how the rating of perceived exertion can be used to determine the maximum pedal load for an electric cart for the elderly. Next, it **explains** the design of a dynamic cart control system that guarantees robust stability. It also **explains** a stability condition based on the concept of dynamic parallel distributed compensation. Finally, it **presents** experimental results that demonstrate the validity of the method.

Typical Mistakes

混用

In this **study**, a two-loop power-flow control system / **was designed**. First, a simple filter that generates a reference current **was devised**. Then, **we presented** a two-loop control system consisting of an inner and an outer loop. …

(Research / Reporting)

「Designed」と「devised」という動詞はこの段落の内容が研究室で行われた仕事についてであることを示す。しかしその次の文にある「presented」は論文の内容を説明している。これは首尾一貫していない。

PRACTICE: 適切な単語で空欄を埋めよ。

1. In this paper, we _____ a wavelength multiplexer …

2. In this work, we _____ a new supercapacitor …

3. This paper _____ a new fabrication technique that …

4. In this research, a new type of signal generator _____.

5. The purpose of this study was to _____ a single-electron device…

6. The purpose of my presentation is to _____ a single-electron device…

novel

「Novel」は賞賛する言葉である。自画自賛しているように聞こえるために、自分の研究にはこの言葉を使うのを避けた方がよい。特に、表題に使うべきではない。他人の研究が画期的ならばそれについては使ってもよい。

ave vs. avg

「Avenue」の略語は「**ave**」である。

「Average」の略語は「**avg**」である。

○ The average power, P_{avg}, was measured.

realize

POINT 1: 英語の「**realize**」と比べると，日本語の「**実現する**」の方が意味は広い。英語には「realize」を使う場面は非常に少ない。

DEFINITION：
1. **デザイン**または**アイデア**をrealizeするとは，物理的な形にすることである。絵を描いたり装置を製作するなどはその例である。
2. **期待，欲求，大望，夢**などをrealizeするとは，それを現実にすることである。

realize + 仮想的なもの・想像上のもの

英語で:
Realize されるものは，具体的なものではなく，**デザイン**や**アイデア**である。

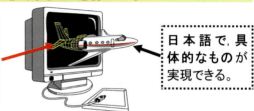

日本語で，具体的なものが実現できる。

「Realize」はおもに将来のことに注目している。

日本語では，「**実現する**」を，通常特定な目標・目的を実現した(する)場合またはタスクを完成した(する)場合に用いる。英語では，「**realize**」はおもに将来のこと，およびアイデア，または概念の実現可能性に注目しているため，特定の研究目標・目的には使わない。また，その過去形「realized」は技術英語ではほとんど使わない。

Good Examples

- If such a system **is ever realized,** it will have many advantages.
  そのシステムは実際に存在しているわけではなく，アイデアだけである。
- **To realize** quantum computing, we need…
  その計算の仕方は現在の時点において実際に存在していない。
- There are several obstacles that have prevented **realization of the full potential** of such a system.
- The results show that these circuits **are realizable** on SOI wafers.
- If devices meeting all these requirements **are realizable**, they can be used in many applications.

realize 5

> **POINT 2:** ただ機械的に「実現する」を「realize」と訳すだけではほとんどの場合間違っていることに注意しよう。

次の動詞を用いて，その使い方(それぞれの動詞にふさわしい目的語)をマスターすれば，作文はさらに正確になりわかりやすくなる。

[動詞:黄色　　目的語:青色]

| achieve | goal, target (目標，目的) |

- This circuit scheme makes it possible to **achieve high speed and low power** at a supply voltage of 0.5 V.
「Achieve」という動詞が使われているから，高速と低パワーが目標であるとわかる。
- These networks require smaller, cheaper modules with a lower loss and a larger number of channels. **To achieve these goals**, we developed …
- A dielectric filter was employed to **achieve good** absolute-wavelength **control**.

| build, make construct | system, network, link, module, package, component, equipment, facility (システム，ネットワーク，リンク，モジュール，パッケージ，コンポーネント，設備，施設) |

- … to **build/construct/make** cost-effective **packages**
- … to **build/construct/make** high-performance **equipment**
- … to **build/construct/make** terabit-class transmission **systems**

| carry out perform | procedure, process, operation, experiment, test, measurement, analysis, simulation, modeling, calculation, research, study, work (プロシージャ，プロセス，オペレーション，実験，テスト，計測，分析，シミュレーション，モデリング，計算，調査，研究，仕事) |

- … to **carry out/perform** a retiming **operation**
- … to **carry out/perform** a numerical **analysis**
- … to **carry out/perform** stress **tests**
- … to **carry out/perform** high-temperature **annealing**

| fabricate | device, circuit, structure, etc. (デバイス，回路，構造など) |

- …to **fabricate/make** integrated **circuits**
- …to **fabricate/make** a **waveguide**
- …to **fabricate/make** a **filter**
- …to **fabricate/make** a **nanostructure**

| implement | function, scheme, logic, design, etc. (ファンクション，スキーム，ロジック，デザインなど) |

- … to **implement** control **logic** (a pseudo-differential **scheme**, a regeneration **function**, an optical Fourier **transform**)

Section 1

obtain — result, value, characteristic（結果，数値，特性）

- A power penalty of less than 1 dB **was obtained** for all wavelengths.
- InGaAsP was used to **obtain** polarization insensitivity.
- Error-free operation **was obtained** on all channels.

provide — desired characteristic, information, service, etc.（望みの特性・特色，情報，サービスなど）

- Portable electronic devices **provide** advanced mobile digital **services**.
- The batteries **provide** a **backup time** of 3 hours.
- This type of antenna **provides** strong **directionality**.

yield — [=To produce as a result.]（「結果として生む」場合に使う）

- This method does not always **yield** accurate results.
- Optimization of the fabrication conditions should **yield** a higher efficiency.
- The data "10110100" were inverted to **yield** "01001011".

PRACTICE: 下の選択肢からできるだけ多くの適切な言葉を選び，次の句を完成させよ。

achieve / fabricate / make / provide
build / construct / implement / yield
carry out / perform / obtain

1. _____ a device
2. _____ a large bandwidth
3. _____ a function
4. _____ a component
5. _____ a digital-to-analog converter
6. _____ planarization
7. _____ a large bandwidth-efficiency product (product =積)
8. _____ multiplexing
9. _____ a system
10. _____ an integrated circuit
11. _____ control
12. _____ annealing
13. _____ an exclusive-or gate
14. _____ accurate results
15. _____ a numerical analysis

confirm

POINT: 英語の「confirm」と比べると，日本語の「確認する」の方が意味が広い。「確認する」を安易に「confirm」に訳してはいけない。

以下の表現を見てみよう。
- 何が起こったかを**確認する**　**to find out** or **determine** what happened

この場合，「確認する」を「confirm」と訳してはいけない。以下は，「confirm」のごく普通の使い方の例である。
- ホテルの予約を**確認する**　**to confirm** a hotel reservation

ホテルの予約を確認するには，前もって予約がされている必要がある。すなわち，**二つのステップ**がある：(1)**まず**ホテルを予約し，(2)**その後**，その予約を確認する。技術英語で，「confirm」を使うには同様に第一ステップが必要である。

DEFINITION: 仮説または記述の正しさを何かにより**confirm**するとは，その仮説または記述が確かに間違いないことを**立証**することである。

NOTE: 「Confirm」を使うには，つぎの二つの要素が必要である。
1. 何かが正しいという**以前の考え（述べたこと）**。
2. その考え（述べたこと）が正しいことを立証する**新たな確証**。この確証により，先の考え（述べたこと）は立証される。

1. First statement

According to my theory, a ray of light passing near the Sun is bent slightly in the direction of the Sun's mass.
（私の理論によると，太陽の近くを通る光線は，太陽の方にわずかに曲げられている。）

2. Confirmation

Our observations **confirm** Einstein's prediction!
（われわれの観測結果はアインシュタインの予言を立証した。）

8   Section 1

Good Examples

- Sato et al. presented the **first experimental evidence** of phosphorus pile-up at the Si/SiO$_2$ interface. ... **Their discovery was confirmed** by sheet resistance measurements...
- Under these conditions, the series resistance **should have** a rather pronounced effect. **To confirm this supposition**, we fabricated circular devices...
- The use of our ICs in fully electrical 40-Gb/s transmission experiments **confirmed our belief** that such systems are both feasible and practical.

結果を報告する場合　（＿＿＿を確認した）

| ~~...was confirmed...~~ ~~We confirmed that...~~ | | ...was found... We found that... |

- The isolation structure (~~was confirmed~~) **was found** to suppress crosstalk.
- It (~~was confirmed~~) **was found** that our scheme reduced the power consumption by over 250 mW.
- (~~We confirmed error-free operation.~~) → We **found** operation to be error-free.

| ~~The results confirmed that...~~ | | The results demonstrate that... |

- These results (~~confirmed~~) **demonstrate** that the device functions properly.
- The evaluation results (~~confirmed~~) **demonstrate** that these modules are suitable for practical use.
- We (~~confirmed~~) **demonstrated** that these techniques will be useful in developing ...

目的を示す場合　（＿＿＿を確認するために）

| ~~To confirm...~~ | | To determine... |

- (~~To confirm~~) **To determine** what effect narrowing the gap has, three structures with different gaps were fabricated.
- The weight of the electrode was measured (~~to confirm~~) **to determine** whether deposition had occurred.
- (~~To confirm~~) **To determine** the maximum throughput, both 10- and 1-Gb/s signals were input.

| ~~To confirm...~~ | | To evaluate/assess/examine... |

- ...(~~to confirm~~) **to evaluate/assess** the effectiveness of the method...
- ...(~~to confirm~~) **to evaluate/assess** the validity of the model...
- ...(~~to confirm~~) **to evaluate** the performance of the search engine...
- ...(~~to confirm~~) **to examine** the feasibility of extending the distance...

that vs. which

That-形容詞節とwhich-形容詞節の違いは，名詞に関する二種類の情報の違いに基づいている。

区別情報： 一つのものと他のものを区別する情報

Ex. 1A:

✗ *The man is a designer.*

三人の男性が絵の中にいるので，「the man」と言う表現は特定の一人と残りの男性を区別するために不十分である。ある特定の人を識別するために，情報を追加する必要がある。

○ The man who is wearing a hat is a designer.

「Who is wearing a hat」と言う表現が，話している人物を**特定するのに必要である**ため，**コンマで区切らない**。

Ex. 2A:

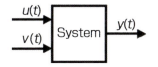

✗ *The input is a sine wave.*

上図には二つの入力があるので，この文は意味を持たない。

○ The input *u(t)* is a sine wave.

「*u(t)*」と言う表現は，一つの入力と他の入力を**区別するために必要である**ので，**コンマで区切らない**。

補足情報： すでに特定されたものに関する補足情報

Ex. 1B:

○ The man is a designer.

絵の中には，一人の男性しかいないので，「the man」と言う表現は特定の男性と他の人たちを区別するのに十分である。さらにもう一つの文を書くことによって，その男に関する補足情報を追加することができる。

○ The man is a designer. He is wearing a hat.

その情報を直接に文の中に入れることもできる。

○ The man, who is wearing a hat, is a designer.

「Who is wearing a hat」と言う表現は，**特定するのに必要でない**補足情報であることを示すために，**コンマで区切られる**。

Ex. 2B:

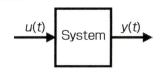

○ The input is a sine wave.

上図には入力が一つしかないので，この文は問題ない。

○ The input, *u(t)*, is a sine wave.

「*u(t)*」と言う表現は，補足情報であるため，**コンマで区切られる**。

that-形容詞節 （区別情報）

この辺にある**全部の店を一つのグループ**と考えてみよう。
○　There are <u>shops in this area</u>.
　That-形容詞節を使って，グループの中にある**特定な店**を**区別して**，話題にしてみよう。例えば，コンピュータを売る店を取り上げたい。

○　There are two <u>shops in this area</u> **<u>that sell computers</u>**.
　That-節を省略すると
✗　*There are two shops in this area.*
となり，コンピュータを売る店がないばかりでなく，この辺には店が二軒しかないという意味になってしまう。そのため，この文は誤りである。このようにthat-節は不可欠であるので，**コンマで区切らない**。

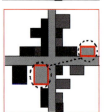

which-形容詞節　(補足情報)

以下の文を見てみよう。
○　The big green shop on the corner was built recently.
「Big green」と「on the corner」という句は一つの店を明確に指す。それに関する**補足情報**(すなわち，その店がコンピュータを売ること)があれば，別の文を追加してその情報を提供する。

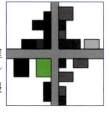

○　The big, green shop on the corner was built recently. It sells computers.
　他の方法として，which-形容詞節を使うことである。
○　The big green shop on the corner**, which sells computers,** was built recently.
　Which-節は店を指すために必要ではないので，**コンマで区切られる**。

> **NOTE:**　形容詞節を省略してみる。それにより文の基本的な意味が変われば，または要点が失われれば，「that」は正しい。そうでなければ，「which」が正しい。

> **Rule 1:That-形容詞節をコンマで区切ってはいけない。**
> **Rule 2:Which-形容詞節をコンマで区切るべきである。**

Good Examples
that

- Portable electronic devices require circuits **<u>that</u>** <u>operate at a low voltage</u>.
- Interest is increasing in e-learning systems **<u>that</u>** <u>adapt to a learner's ability</u>.
- The goal is to find an optimal scheme **<u>that</u>** <u>minimizes production costs</u> .
- A mechanism **<u>that</u>** <u>automatically selects a suitable pedal load</u> was developed.

Adjective Clause: Short Form   11

which

- India ink contains carbon particles**, which** are electrically conductive.
  - (= India ink contains carbon particles. They are electrically conductive.)
- The output of a solar panel depends on the amount of solar radiation**, which** is affected by the cloud cover（雲量）.
  - (= The output of a solar panel depends on the amount of solar radiation. That amount is affected by the cloud cover.)
- Several testing rigs use heat pumps**, which** can heat and cool the circulating fluid.
  - (= Several testing rigs use heat pumps. A heat pump can heat and cool the circulating fluid.)
- V_m is the voltage applied to the motor**, which** is proportional to the driving torque.
  - (= V_m is the voltage applied to the motor. It is proportional to the driving torque.)

Adjective Clause: Short Form

この文法は技術英語を理解するために、また、優れた英文を書くために不可欠である。

That

POINT:「That」は形容詞節の主語である場合、しばしば削除してもよい。

Pattern 1:   *that* + 能動態の動詞   ⇒   現在分詞

1. Sample 1 is a reference **that consists** of pure polyethylene.
   ⇒ Sample 1 is a reference **consisting** of pure polyethylene.
2. A reference **that consisted** of pure polyethylene was measured first.
   ⇒ A reference **consisting** of pure polyethylene was measured first.
   動詞の時制は無関係であることに注意しよう。現在分詞の形はいつも同じである。
3. Intermolecular interactions **that involve** hydrogen bonds help to stabilize the crystalline form.
   ⇒ Intermolecular interactions **involving** hydrogen bonds help to stabilize the crystalline form.

Pattern 2:   *that* + be-動詞   ⇒   that + be-動詞を削除する

1. The module contains two components **that were** written in JAVA.
   ⇒ The module contains two components written in JAVA.
2. The 2D electron gas has a response **that is** similar to that of the coil.
   ⇒ The 2D electron gas has a response similar to that of the coil.
3. The membrane is a rectangular plate **that is** 120 μm wide and 600 μm long.
   ⇒ The membrane is a rectangular plate 120 μm wide and 600 μm long.

12   Section 1

「That」+be-動詞を削除した後，残りの表現が**非常に簡単**であれば，それを名詞の前におくこともある。

- ○ The characteristics of <u>devices that were 3 μm long</u> were measured.
  - ⇨ ○ The characteristics of <u>devices 3 μm long</u> were measured.
  - ⇨ ○ The characteristics of <u>3-μm-long devices</u> were measured.

- ○ The <u>results **that were** obtained</u> agree well with the calculation results.
  - ⇨ ○ The <u>results obtained</u> agree well with the calculation results.
  - ⇨ ○ The <u>obtained results</u> agree well with the calculation results.

DANGER

残っている表現が簡単でない場合，それを分けて，一部を名詞の前におき残りの部分を名詞の後ろにおいては絶対にいけない。

- ○ The results **that were** obtained by our new method agree well with the calculation results.
  - ⇨ ○ The results obtained by our new method agree well with the calculation results.
  - ⇨ ✗ *The obtained results by our new method agree well with the calculation results.*

  「Obtained」と「by」との間の関係が成立しなくなる。ここで，「by」の句が「results」を修飾しているが，「results by our new method」という表現に意味はない。

- ○ Devices **that were** fabricated by our new technology were examined.
  - ⇨ ○ Devices fabricated by our new technology were examined.
  - ⇨ ✗ *Fabricated devices by our new technology were examined.*

  「Fabricated」と「by our new technology」との間に関係はないので，この文に意味はない。

PRACTICE: 以下の形容詞節を短い形に変えよ。

1. Measurements yielded results <u>that are similar to those for GaMnAs</u>.

    _____

2. The optical power <u>that comes from the two ports</u> is 8% of the output power of the laser.

    _____

3. The reliability of heterojunction bipolar transistors with an emitter <u>that was 0.6 μm wide and 3 μm long</u> was examined.

    _____

Adjective Clause: Short Form   13

4. A mask structure <u>that consists of thin and thick layers</u> was used for the fabrication.

5. The power of the light <u>that passes through the fibers</u> is monitored.

6. The voltage <u>that was applied to the device</u> varied from 1 V to 5 V.

7. A piece of concrete <u>that contained a 0.35-mm-wide crack</u> was used in the experiment.

8. The current <u>that is induced by the signal</u> flows through the coil.

9. The angle of rotation depends on the voltage <u>that is applied to the electrode of the mirror</u>.

10. A receiver <u>that operates at a frequency of 250 GHz</u> was fabricated and tested.

11. A sensor <u>that is mounted on the tip of an optical fiber</u> was used for the measurements.

12. The carriers (キャリア) <u>that are injected into the device</u> attenuate the optical power <u>that reaches the output fiber</u>.

Which

POINT 1: That-形容詞節と同じように,「**which**」が形容詞節の**主語**であれば, それを削除できる場合もある。
POINT 2: コンマは必要である。

1. A lower capacity, **which indicates** a change in properties, was observed.
   ⇨ A lower capacity, **indicating** a change in properties, was observed.

2. This procedure selects a small region, **which is called** the subwindow.
   ⇨ This procedure selects a small region, **called** the subwindow.

Section 1

Changes & Differences

POINT 1: 以下のパターンはどの種類の変化を示す**名詞**に対しても使ってよい: change, increase, decrease, variation, fluctuations, reduction, rise, drop, degradation, improvement, etc.

Basic Pattern

a change in voltage of 5 V

「**In**」は変化する**状態量**を示す。

「**Of**」はその**変化の度合い**を示す。

数値と単位の間に**スペースを入れる**。

- a decrease in power of 1 dB
- a variation in pulse width of 6 ps
- fluctuations in sensitivity of 0.5 dB
- a drop in β of 10%

Other Patterns

名詞の前の単位と数値との間のスペースが**ハイフン(-)**に変わる。

a 5-V change in voltage

- a sharp rise in temperature
- the gradual increase in the gain
- the variation in width
- the degradation in quality
- fluctuations in the voltage
- a 2-dB improvement in the SNR
- a 10% drop in β
- a 5-μA increase in the current
- a 13-nm variation in width
- a 3-dB reduction in noise

a voltage change of 5 V

- a temperature increase
- amplitude fluctuations
- a sudden voltage drop
- the power variation
- a flow-rate variation of 5%
- a temperature drop of 17°C
- sensitivity fluctuations of 0.5 dB

POINT 2: 「**Difference**」の使い方は以上の「change」のパターンとほぼ同じである。違いは「between」も使えることだけである。

Valley 15°C
Coast 25°C

Basic Pattern

a difference in temperature of 10°C between the valley and the coast

Changes & Differences 15

Between: There is a **difference between** X and Y.

- There is a **difference between** the valley and the coast.
- There is a **difference between** <u>the temperature of</u> the valley and <u>that of</u> the coast.
- There is a **difference between** <u>the temperatures of</u> the valley and the coast.
  温度が二つあるから，複数形の「temperatures」が正しい。

Other Patterns

- There is a **difference in** temperature.
- There is a **difference in** temperature **between** the valley and the coast.
- There is a 10°C **difference in** temperature **between** the valley and the coast.

- There is a **difference in** temperature **of** 10°C.

- There is a **difference of** 10°C.
- There is a **difference of** 10°C **between** the valley and the coast.
- There is a **temperature difference of** 10°C **between** the valley and the coast.

PRACTICE: 「In」か「of」か「between」を使って，次の空欄を埋めよ。

1. There is a 10-dB difference _____ noise level _____ these two places.
2. The use of flip-chip interconnects yields a reduction _____ crosstalk _____ about 5 dB.
3. An increase _____ the amount of moisture in dry soil can lead to a great increase _____ thermal conductivity.
4. There is a difference _____ output power _____ less than 1.0 dB _____ channels.
5. We obtained a difference _____ matching rate _____ 0.05.
6. Changes _____ the speed of rotation of a wind turbine cause fluctuations _____ the frequency and voltage of the output.
7. The refractive-index difference _____ the core and the cladding is 0.4%.
8. A reduction _____ onset gain from 4.8 to 2.8 is equivalent to an increase _____ the high-power tolerance _____ over 2 dB.

9. The results were obtained for power-supply fluctuations _____ 5%.
10. This difference _____ characteristics _____ the two samples could be due to a difference _____ current-blocking properties.
11. A 20% drop _____ the peak-to-peak voltage was simulated.
12. A 3-dB improvement _____ the signal-to-noise ratio and a 3-dB reduction _____ noise were obtained.
13. The difference _____ the free energies of Reactions A and B is very small.
14. There is a small difference _____ free energy _____ Reactions A and B.

reduction **in** CO_2 emissions   vs.   reduction **of** CO_2 emissions

変化を示す単語の一部は二つの意味を持っている。

度合い: 状態量変化の度合いを示す場合，単語の後ろの前置詞は「in」を使う。
- Our goal is a 10% **reduction in** CO_2 emissions within two years.

行動: 行動を示す場合，単語の後ろの前置詞は「of」である。
- The **reduction of** CO_2 emissions is the main focus of our green-energy policy.
  (= Reducing CO_2 emissions is the main focus…)

両方の例:
- The **reduction of** operating costs results in an **increase in** profits.

> **POINT 3:** 上記の議論は，変化を示す名詞に関連する前置詞についてである。変化を示す**動詞**の場合は，前置詞は異なる。

Basic Pattern

> *Y* changes by *ΔY* from Y_0 to Y_1.

- The phase fluctuates by a small amount.
- Aging causes the maximum heart rate to decrease by 0.685 beats/minute/year.
- The voltage gradually increases to V_{DD}.
- The inverted-magnetic-field region broadens as the temperature decreases from 44 K to 25 K.
- As the thickness changes from 0.45 mm to 0.55 mm, the center frequency changes from 125 GHz to 115 GHz.

first vs. at first

POINT: 「At first」は誤って，「first」の意味で使われることが多い。

「First」は一連またはリストの項目に使用される。

First	Next	Then	After that	Finally
First	Second	Third	Fourth	Last

NOTE:「Firstly」，「secondly」などは昔風の英語である。

Good Examples

- Figure 3 illustrates the fabrication process. **First**, Cu interconnection patterns are formed. **Next**, Al film is deposited. **Then**, annealing is carried out. **After that**, the Al on the SiO_2 is selectively removed. **Finally**, the sample is heated to 350°C.
- This paper **first** explains the current modulation mechanism. **Next**, some experimental results are presented to show that...
- Oxygen species affect the composition through various kinetic routes. **First**, oxygen atoms coming from… **Second**, oxygen atoms can be directly incorporated… **Third**, activated oxygen species react...
- However, there are problems with this configuration. **The first** is the requirement for a more compact system. **The second** problem is the need to prevent dust and eye injuries. **The third** concern is reproducibility.

「At first」は変化する状況を描写するのに使われる。

At first, it was sunny; but **later**, it got cloudy.

技術英語では，「at first」を使う場面が少ない。

Good Examples

- We start with a gate voltage of zero, and then increase it. <u>**At first**</u>, there are no excess electrons in the island. <u>**But** when the gate voltage crosses the dotted line</u>, one electron tunnels through into the island.

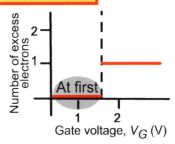

18   Section 1

- This figure illustrates how cooling affects the stress in a film. The stress does not increase very much **at first** because the generation of dislocations relieves it. **But** below 400°C, it rises rapidly.

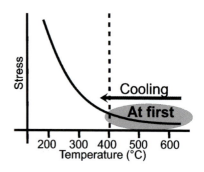

operating principle

POINT: 名詞の前には,「operation」ではなく「operating」を使用すべきである。

~~operation principle~~ ➡ operating principle

- Section 2 explains the ~~(operation)~~ **operating principle** of the device.
- The ~~(operation)~~ **operating voltage** is 3.3 V.
- The performance of the driver amplifier limits the ~~(operation)~~ **operating speed**.

PRACTICE:「Operation」または「operating」を選択し空欄を埋めよ。

1. For pulsed _____, the threshold current is 19 mA.

2. The maximum _____ frequency is 39 GHz.

3. Some lasers have been developed for high-temperature _____.

4. Figure 7 shows the transient response when the _____ mode changes.

5. The results show that a 1-by-43 switching _____ is possible.

6. It is important to reduce _____ costs.

7. The equipment must function within the required _____ temperature range.

8. Error-free _____ was obtained at a data rate of 400 Gb/s.

evaluate vs. estimate

DEFINITION: 何かを **evaluate** するとは，よい点と悪い点を検討し，その重要さ，価値，または質を決めることである。

POINT 1: Evaluate の結果はよいか悪いか，適切か不適切か，ふさわしいかふさわしくないか，十分か不十分などであり，**数量ではない**。

- A test device was fabricated in order to **evaluate its performance**.
- **The effectiveness** of the new method **was evaluated**.

DEFINITION: 数量を **estimate** するとは，大まかな計算・計測をすることである。

POINT 2: Estimate の結果は**数量**である。

- The length **was estimated** to be about **0.2 μm**.
- The capacitance **can be estimated** from the period of the voltage oscillations.

Typical Mistakes

- The change in the refractive index can be (~~evaluated~~) **estimated** from the amplitude of the oscillations.
- The controller (~~evaluates~~) **calculates** the matching ratio.
- A simple method of (~~evaluating~~) **determining/estimating** the height of a silicon structure after oxidation was devised.
- The electrical characteristics were (~~evaluated~~) **measured** at room temperature .

PRACTICE A: 「Evaluate」か「estimate」または両方を使って，次の空欄を埋めよ。

1. _____ the crystal quality
2. _____ the defect density
3. _____ the width
4. _____ the switching time
5. _____ the reliability
6. _____ the method
7. _____ the usefulness of the technique
8. _____ the difference in length
9. _____ the variation in size
10. _____ the device performance
11. _____ the time required

Section 1

POINT 3: 「評価する」の英訳はいくつかある。ただ機械的に「**evaluate**」と訳すと，間違っている可能性は高い。

PRACTICE B: 次の文に「evaluate」を使用するのは適切ではない。「Evaluate」の代わりに下記の選択肢から適切な単語を選択し空欄を埋めよ。

**determine   estimate   examine   identify
investigate   measure   observe**

1. The depth of the holes was _____ from microscope images.
2. The contaminants (汚染物質) were _____ by thermal desorption spectroscopy (熱脱着スペクトロスコピー).
3. The structure of the polysilicon was _____ from microscope images.
4. To _____ the effect of applying an RF bias during the deposition of WSiN, the surface composition of the resulting film was analyzed.
5. The uniformity of the patterns was _____ by measuring the intensity of diffracted light.
6. Atomic force microscopy (原子間力顕微鏡法) is now widely used to _____ surface structures.
7. We _____ the dependence of current on voltage.

NOTE: 「Estimate」という動詞に二つの名詞が関連している：「estimation」と「estimate」。

DEFINITION: Estim<u>ation</u> は大まかな計算・計測をする動作(action)である。

- **Estimation** of the dopant concentration requires precise measurement of the resistance. (= In order to estimate the dopant concentration, you have to measure the resistance precisely.)

DEFINITION: Estimate は大まかな計算・計測をする結果，すなわち数量である。

「**Estimate**」の発音：　　動詞: es′tə・māt　　名詞: es′tə・mit

- **Our estimate of** the density differs significantly from previously reported values. (= The value we obtained for the density is different from what other researchers obtained.)
- **An estimate of** the disturbance is incorporated into the control input to improve the control performance.

Units

Abbreviation vs. Full Word

数の後または図やチャートの中では単位の略語を書くこと。
- The lasing wavelength was 1.5 μm.

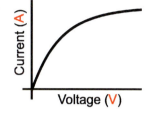

上のケース以外は，単位を普通の単語として書くこと。
- …a (~~mm-wave~~) millimeter-wave signal…
- …(~~THz~~) terahertz imaging…
- Plasma effects appear within (~~a few ns~~) a few nanoseconds.

Abbreviation vs. Symbol

一番よく使う記号:　° ' " % # ¥ $ ¢ £

数値＋単位 (Space)	数値＋単位略語 (Space)	数値＋記号 (No space)
英語では数を単語として扱うため，数とその次の単語の間にスペースを入れること。	英語では略語を単語として扱うため，数とその次の略語の間にスペースを入れること。	英語では記号は単語として扱わないため，数とその次の記号の間にはスペースは入れないこと。
7 degrees Celsius	7 deg Celsius	7°C
6 inches	6 in.	6"
96 percent		96%
24 nanometers	24 nm	
1.4 Volts	1.4 V	
8 micrometers	8 μm	
2.4 gigahertz	2.4 GHz	
23 dollars and 50 cents		$23.50

Space vs. Hyphen

名詞の前の形容詞句として使わない場合，数と単位の間にスペースを入れること。
- Measurements were made at a voltage of 5 V.
- The devices were 3 μm long.

名詞の前の形容詞句として使う数と単位の間に**ハイフン**を入れること。
- Measurements were made on 3-μm-long devices.

名詞の後で，短い形の形容詞節としての数と単位の間には，ハイフンではなく，**スペース**を入れること。
- Devices 3 μm long were fabricated. (= Devices that were 3 μm long…)

In A Graph

図の**軸のラベル**で，単位を略語ではなく普通の言葉として書くとき，**複数形**を書くこと。

Arbitrary units (任意単位)：図の中で，「arbitrary unit(s)」という表現を使いたい場合，単数形の「arb. unit」と比べて，複数形の「**arb. units**」または略語の「a.u.」の方がよく使われている。

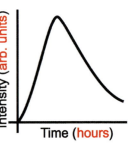

Period vs. No Period

この文を見てみよう。
- The wafer is 6 in in diameter.

「In in」は間違えに見えるから，この文は読みにくい。「In」は普通の英語の単語であるから，「in.」という「inches」の略語にピリオドが必要である。
- The wafer is 6 in. in diameter.

単位の略語は英語の単語ではない場合 (cm, μA, pf, Gb, ns, など)，ピリオドを書く必要はない。

enable

POINT:「Enable」を「**be　able　to**」と混同してはいけない。文法が違う。「enable」は**他動詞**であるので，目的語は必要である。

　　NOT ENGLISH!

Pattern 1: enable X
- This technique **enables** the **fabrication** of walls with a high aspect ratio.
- Si photonics **enables** high-density **integration** on a Si wafer.

Pattern 2: enable X to do Y
- A large bandwidth **enables** wireless **systems to handle** multigigabit data streams.
- This alignment method **enables** the **cores to be aligned** directly.

Space

1. 数値と単位の略語との間にスペースを一つ入れること。

- ✗ *The supply voltage was 5V.*
- ○ The supply voltage was **5 V**.

• … at a frequency of (~~40GHz~~) **40 GHz**

• The device is (~~50μm~~) **50 μm** long.

1 mm

NOTE: 「数値＋単位」を名詞の前におく場合，数値と単位との間のスペースはハイフンに変わる。

- …a **40‑GHz** optical signal…
- …a **50‑μm‑long** device…

2. 数値と単位の記号との間にはスペースを入れないこと。

一番よく使う記号: ° ' " % # ¥ $ ¢ £ 550°C 55% 6" $23.50 45°

3. 括弧の前後にスペースを一つずつ入れること。

- ✗ *Extreme ultraviolet lithography(EUVL) is a promising way to …*
- ○ Extreme ultraviolet lithography **(EUVL)** is a promising way to …

- ✗ *… silicon-on-insulator(SOI) technology …*
- ○ … silicon-on-insulator **(SOI)** technology …

- ✗ *A large λ will satisfy all the timing constraints(if they can be satisfied)but will result in poor values of W.*
- ○ A large λ will satisfy all the timing constraints (if they can be satisfied) but will result in poor values of *W*.

4. 数式の等号または不等号の前後にスペースを一つずつ入れること。

- ✗ *When β=1, …*   ✗ *When β<1, …*
- ○ When $\beta = 1$, …   ○ When $\beta < 1$, …

5. グラフの軸の変数名と単位の間にスペースを一つ入れること。

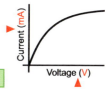

6. ラベル名と数字の間にスペースを一つ入れること。

- ✗ *Figure3*   ✗ *Fig.3*   ✗ *Eq.(5)*
- ○ Figure 3   ○ Fig. 3   ○ Eq. (5)

図中の部分を参照するために様々なスタイルがあって，どんなスタイルが一番いいかという決まりはない。以下に三つの一般的なスタイルを示す。

スペースを一つだけ入れる	コンマの後ろにもスペースを入れる	スペースも括弧も使う
● In Fig. 3a, it can be seen that… ● In Figs. 3a-b,d,4b,d, it can be seen that…	● In Fig. 3a, it can be seen that... ● In Figs. 3a-b, d, 4b, d, it can be seen that…	● In Fig. 3(a), it can be seen that… ● In Figs. 3(a)-(b), (d), 4(b), (d), it can be seen that…

7. 章立て・リストの数値あるいは文字の後ろにスペースを一つか二つ入れること。

× 1.Introduction
○ 1. Introduction

8. 参考文献の関連箇所に適切にスペースを入れること。

4. H. Li, C. White, and R. D. Scott, "A 3-GHz Wave-Pipelined Adder," *IEEE Journal of Solid State Circuits*, Vol. 68, No. 5, pp. 517-529, Sept. 1994.

9. ガイドライン：字下げスタイルの段落において，字下げの目安としては少なくとも半角スペース三つ（＝日本語ワープロの1.5字）である。

POINT 1: 多くのジャーナルは，インターネットからダウンロード可能なテンプレートを用意してあり，それを使うべきである。それがない場合，インターネット上のスタイルマニュアルから，正しい字下げスタイルを確認する。それもない場合，上記のガイドラインに従う。
POINT 2: 段落の先頭で字下げする場合は，三つ以上の半角スペース（英文スペース）を使うべきである。一つか二つだけでは，スペースが少なすぎるため，間違いだと思われる。

10. 英語の論文に日本語のスペースを使わないこと。

　MS Word を使う場合には，スペースが表示できる：Wordの画面から，ファイル→オプション→表示→常に画面に表示する編集記号→「スペース」をオンにすれば，英文スペースと日本語のスペースが簡単に区別できる。

- 英語のスペース： ・ （小さいドット）
- 日本語のスペース： □ （正方形）

× □There・are・three□Japanese・spaces・in・this□sentence.

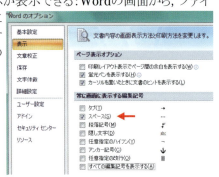

Dynamic Verbs 1

動詞は名詞よりパワフルである。

次のペアとなっている文を吟味しよう。

A1: Argon implantation <u>causes</u> membrane <u>deformation</u>.
A2: Argon implantation **deforms** the membrane.

B1: Electrolyte <u>decomposition occurs</u> continuously.
B2: The electrolyte continuously **decomposes**.

各ペアとも二番目の構文の方がはるかによい。「Deform」と「decompose」という動詞の使用により，文がダイナミックになっている。また，名詞の代わりに動詞を使うことにより文は（短くなり）理解しやすくなる。

PRACTICE: 太字の名詞を動詞に変えよ。

1. Hot-electron **injection** <u>was carried out on</u> the transistors.

   Hot electrons _____.

2. There is a bandwidth **<u>limitation</u>** <u>due to</u> the electrical interface.

   The electrical interface _____.

3. Figure 3 <u>shows an</u> **illustration** <u>of</u> the experimental setup.

   Figure 3 _____.

4. Polymer layer **removal** <u>was performed</u>.

   The polymer layer _____.

5. <u>The</u> **organization** <u>of</u> this paper <u>is</u> as follows:

   This paper _____:

6. <u>The</u> **measurement** <u>of</u> displacement can be accurately <u>done</u> with an encoder.

   The displacement _____.

7. It is difficult to achieve a dramatic **improvement** in the performance.

    It is difficult to _____.

8. Voltage drops in a utility grid can cause severe **damage** to sensitive loads.

    Voltage drops in a utility grid _____.

9. A wavelength **conversion** operation from λ_1 to λ_2 is accomplished.

    The wavelength _____.

10. The **design** of K_P and $F(s)$ can be carried out independently.

    K_P and $F(s)$ _____.

Prepositions 1

空欄を適切な前置詞で埋めよ。必要がなければ、×で埋めよ。

CHECK YOUR KNOWLEDGE

1. This paper discusses _____ a new approach...

2. X equals _____ Y.

3. X is equal _____ Y.

4. X is the same _____ Y.

5. X is due _____ Y.

6. X consists _____ Y and Z.

7. X influences _____ Y.

8. This section explains _____ the method.

9. X is called _____ Y.

10. X depends _____ Y.

Section 2

propose

Lists

Specifying Values

depend on

contain vs. include

on the contrary

adopt

cannot   can not   can't

in case of fire

Connecting Nouns

apply

Punctuation: Hyphen (-)

Style: Dynamic Verbs 2

Prepositions 2

propose

POINT 1: 「Propose」という単語は科学界に新しいアイデアを提案する、ということである。Proposeされたアイデアを、他の研究者は自由に使用できる。単に新しいものを示し、または新しいものを製作したことを報告する場合には、「propose」は使わない。

DEFINITION: 計画、アイデア、手法などを propose するとは、他の人がそれを検討したり、使ってみたり、決めたりできるように、それを提案することである。

「Propose」は一般的に二つの場合に使用される。

基礎研究：例えば、新しい理論的な概念または新しい手法を、他の研究者にテスト・検証してもらうために、proposeする。

標　　準：例えば、新しい標準を標準化委員会にproposeする。

Good Examples

- H. A. Lorentz **proposed** the electron theory of electrical charge in 1895; and in 1897, J. J. Thomson of England showed that electricity was indeed caused by negatively charged particles (electrons).

- Alec J. Jeffreys first **proposed** DNA analysis in 1985.

- In this paper, we **propose** a new concept for ultrafast digital ICs made with traveling-wave FETs called traveling-wave FET logic.

- Since the mean opinion score for harmonic vector excitation coding (HVXC) was the best among all the **proposed** coders, HVXC was chosen to be the ISO/IEC International Standard for MPEG-4 Audio.
  MPEG-4オーディオ標準の候補としてコーダーは提案され、標準化委員会はどれが一番よいかを決定した。

多くの研究者にとって「propose」を使う場面は少ない。

POINT 2: 「Proposed device」は通常実在しないものである。デバイスをproposeするとは、ただ単にその概念を示し、おそらく正しく稼動すると主張することである。この場合、デバイスはまだアイデアの段階で、実現されていないという意味合いが強い。

propose   29

問題を解決する，または目標を達成することを表現するには，新しいものを **develop** する，または **devise** するという表現を用いることを薦める。単に新しいものを製作した事実を報告する場合，「propose」を使わないことにしよう。

~~propose~~ ➡ **develop, devise**

- Coupling to a fiber is a serious problem with Si wire waveguides. To solve it, we (~~proposed~~) **developed/devised** a spot-size converter.
  デバイスをproposeするとは，単にその概念を提案することである。上の文を書いた研究者はデバイスをすでに製作して実験的検証も行ったので，「propose」を使うのは間違いである。
- We (~~proposed~~) **developed/devised** a new circuit configuration to achieve high speed.
- We (~~proposed~~) **have developed/devised** a chip-size cavity package that...

新しいデバイス，手法などについて「the proposed X」としないこと。特に，Point 2で説明したように，「**proposed device**」とは実在しないデバイスであることに注意しよう。

~~proposed~~ ➡ **[Nothing], new, fabricated**

- This graph shows the power dissipation of (~~the proposed~~) **the** circuit (**our new** circuit, **a fabricated** circuit).
  Proposed 回路はまだ実現されていないので，パワーの消散は計測できないはずである。
- For the (~~proposed~~) **new** etching process, the etching rate was 389 nm/min.
- (~~The proposed~~) **This** interconnection provides a low connection loss. (**The new** interconnection…)

PRACTICE: 以下の文では，「propose」または「proposed」が削除されている。代わりにもっとも適切な言葉を空欄に入れよ。

1. To solve the problem, we _____ a new circuit.
2. This shows the structure of _____ transistor.
3. To obtain good performance, we _____ new hardware and software architectures.
4. This paper _____ a dynamic flip-flop circuit with a low power consumption.
5. We _____ a new type of sensor.
6. The purpose of this study was to _____ a 3-D hardware algorithm.

Lists

コロン(:)の有無

コロンをつける場合

リストの前のテキストが完成文になっていれば、コロンをつけること。
- There are three requirements to be considered in designing the device.

以上のテキストは完成文なので、リストを追加するとき、**ピリオドをコロンに変る**こと。
- There are three requirements to be considered in designing the device: shock resistance, compactness, and reliability.

通常、コロンを使う場合、リストは文の最後にあり、その後ろにピリオドがある。すなわち、リストの後ろに他のテキストを追加しないこと。
- ✗ *Three variables were measured: length, mass, and temperature, and a statistical analysis was performed.*
- ○ *Three variables were measured: length, mass, and temperature. Then, a statistical analysis was performed.*

コロンをつけない場合

リストの前のテキストは完成文になっていない場合、コロンをつけないこと。
- The requirements to be considered in designing the device are

以上のテキストは完成文ではないので、リストを追加するとき、コロンをつけないこと。
- The requirements to be considered in designing the device are shock resistance, compactness, and reliability.

パラレル構造

項目は文法的な構造は同様であること。
- ✗ *The requirements to be considered in designing the device are*
    - ◇ *to withstand shocks,*　　不定詞
    - ◇ *compactness, and*　　名詞
    - ◇ *very reliable.*　　形容詞
- ○ The requirements to be considered in designing the device are
    - ◇ shock resistance,
    - ◇ compactness, and
    - ◇ reliability.

インライン(行内)リスト

完全リスト　（項目全部がリストアップされる）

- The variables that were measured are length, mass, and temperature.

Lists   31

- Three variables were measured**:** length**,** mass**,** and temperature**.**
- Three variables **(**length**,** mass**,** temperature**)** were measured.
  リストは括弧で囲っている場合、「and」を使う必要はない。
- Three variables—length**,** mass**,** and temperature—were measured.
- Three variables**,** **namely,** length**,** mass**,** and temperature**,** were measured.

不完全リスト　（一部の項目だけがリストアップされる）

such as　　including　　for example　　e.g.

　この表現自体はリストの一部だけがリストアップされることを示すので、「etc.」とか「and so on」などをつける必要はない。
- Many variables**,** **such as** length**,** mass**,** and temperature**,** were measured.
- Many variables**,** **including** length**,** mass**,** and temperature**,** were measured.
- Many variables**,** **for example,** length**,** mass**,** and temperature**,** were measured.
- Many variables**,** **e.g.,** length**,** mass**,** and temperature**,** were measured.

etc.

- Many variables were measured**:** length**,** mass**,** temperature**, etc.**
  a. 「Etc.」が文の最後の単語である場合、ピリオドを一つだけ使うこと。
  b. 「Etc.」が「…とその他」を意味する「et cetera」の略語である。「Et」がラテン語の「and」であるから、「and etc.」は間違いである。
- Many variables (length**,** mass**,** temperature**, etc.**) were measured.

and so on

「And so on」は「etc.」と少し違う。想定する読者が主題をよく知っている時は、リストを簡単に続けることができる場合、「and so on」を使うこと。
- There are several large cities in Japan—Tokyo, Osaka, Fukuoka**, and so on.**
  日本人は日本の大きな都市をよく知っているので、このリストを簡単に続けることができるため、この文章は日本人の読者に適切である。しかし、外国人には、日本の都市を知らない可能性が高いので、この文章は適切ではない。それらの人々のために、代わりとして、「etc.」を使用した方がよい。
- Many variables—length**,** mass**,** temperature**, and so on**—were measured.

X, Y, Z, and other ABCs

以下の文を見てみよう。
- Active structural control is an effective way to protect **structures, such as buildings, towers, and bridges,** from vibrations due to strong winds or earthquakes.

Section 2

この文が正しいとしても，リストは文の自然な流れを妨げている。以下のようにすることによって，文の流れがスムーズになる。

- Active structural control is an effective way to protect **buildings, towers, bridges, and other structures** from vibrations due to strong winds or earthquakes.

> **PRACTICE:** 以下の文の中のリストを「and other...」形式に変化せよ。
> Ex. : This type of neural network provides **benefits, such as** a simple structure and fast learning.
> ⇨ This type of neural network provides a simple structure, fast learning, **and other benefits**.

1. This modulator provides the required characteristics, for example, high speed and a low driving voltage.

2. Intelligent control methods, for example, neural networks and adaptive fuzzy control, have been used to deal with this problem.

3. The acceptable crack width depends on ambient conditions, such as humidity.

4. The components are affected in different ways by factors such as aging, voltage fluctuations, and variations in temperature.

5. The report covers technical and operational issues, for example, propagation problems, system design parameters, and possible applications.

6. Nonlinearities, e.g., a dead zone and hysteresis, can seriously degrade control performance.

7. There is little space for optical devices, such as an isolator and filters.

Lists   33

一項目一行のリスト

以下の文で，x_mは何を意味するのかを確認してみよう。

- In the model of the dual-stage feed drive (Fig. 1), the parameters f_m, f_M, m, M, c_m, k_m, x_m, x_M, and y_P are the input force produced by Actuator 1, the input force produced by Actuator 2, the mass of the fine stage, the mass of the coarse stage, the damping factor of the fine stage, the stiffness of the fine stage, the distance of the fine stage from the coarse stage, the distance of the coarse stage from the reference point, and the distance of the fine stage from the reference point, respectively.

とても読みにくい！文法は正しいが，情報提示の仕方は適切ではない。以下の文のように，項目はペアであるか，または複雑であれば，一項目ずつ違う行で記述することを考えよ。なお，項目の中に一つでも完成文を含めば，すべての項目を別の行で記述すること。

- The parameters of the dual-stage feed drive (Fig. 1) are
  - f_m input force produced by Actuator 1
  - f_M input force produced by Actuator 2
  - m mass of fine stage
  - M mass of coarse stage
  - c_m damping factor of fine stage
  - k_m stiffness of fine stage
  - x_m distance of fine stage from coarse stage
  - x_M distance of coarse stage from reference point
  - y_P distance of fine stage from reference point.

以上の文で，x_mの意味はわかりやすい。読者のことを考えて，情報を理解しやすく提示すること。

リストの句読法： 項目は単語または句である場合

様々な一項目一行のリストの句読法のスタイルがあって，どんなスタイルが一番いいかという決まりはない。以下に二つの一般的なスタイルを示す。

Style 1: 普通の文としてリストに句読点をつけること。

The requirements to be considered in designing the device are
- shock resistance,
- compactness, and
- reliability.

Style 2: ただ最後の項目の後ろにピリオドをつけること。

The requirements to be considered in designing the device are
- shock resistance
- compactness
- reliability.

リストの句読法: 項目は完成文である場合

大文字で始まり，ピリオドまたは疑問符で終わること。
There are three requirements to be considered in designing the device**:**
- ◇ The device must be shock resistant, compact, and reliable**.**
- ◇ The device must meet international standards**.**
- ◇ The device must have a battery life of at least 72 hours under conditions of normal use**.**

セミコロン（;）

コンマを含む項目があれば，**セミコロン**ですべての項目を分割すること。
- Three variables were measured: length**,** L**;** mass**,** m**;** and temperature**,** T.
- Three variables (length**,** L**;** mass**,** m**;** temperature**,** T) were measured.
- The generator consists of two fixed-wavelength laser sources**,** LS1 and LS2**;** a tunable laser**;** two optical couplers**;** and two photomixers.
- The requirements for using Si photonic devices in telecom applications are very severe: low insertion and polarization-dependent losses**;** fast active functions**,** such as optical modulation and reception**;** good reliability**;** low cost**;** etc.

アルゴリズム ＆ 手順

アルゴリズムまたは手順の正確なステップをリストアップするときには，命令法を使用すること。
- Procedure for building an SD neural-network model:
    1. Select 500 groups of samples...
    2. Choose an $R_p(i)$ in the range [0.2, 0.5] for...
    3. Let $\{\varphi_1, \varphi_2, ..., \varphi_c\}$ be the initial set of...
    4. Calculate the Euclidean distance between...
        -
        -
        -

depend on   35

Specifying Values

POINT: 温度や電圧などの変数の数値を記す場合，一番よく使用される方法は以下のパターンである。

BASIC PATTERN: (a/an)　変数　前置詞　数値

単数名詞: **a temperature of** 325°C
複数名詞: ~~the~~ **temperatures of** 240°C and 325°C

- …a transistor with **a** gate length **of** 0.25 μm…
- The stage moves through **a** distance **of** 125 mm.
- Figure 3 shows simulation results for **a** voltage **of** 2 V.
- This receiver operates at data rates **of** over 10 Gb/s.
  上の文において，「over」は形容詞であり，前置詞ではないことに気をつけよう。下線の引かれた部分は data rates of more than 10 Gb/s と同じである。
- The main peaks are at frequencies **between** 1.1 THz and 1.3 THz.
- Growth temperatures **above** 500°C enhance the desorption of…
- Drain voltages **of** 0.1 V and 2 V were used for the nMOSFETs.

このパターンは**記号**にも適用できる。
- Measurements were made at **a** θ **of** 23°.
- The power dissipation is 1.3 W at **a** V_{DD} **of** 3.3 V.
- Simulation results for βs *of* 0.32 and 0.48 show that…

NOTE:「A 325°C temperature」とか「a 5-V voltage」などのように，数値を状態量の前におくパターンはほとんど使われない。

depend on

POINT:「Is depend on」は正しい英語の表現ではない。

- ○ X **depends on** Y.
- ○ X **is dependent on/upon** Y.

✗ ~~X is depend on~~ Y.
NOT ENGLISH!

「Be dependent on/upon」と比べて，「depend on」の方はよく使われている。
- The wavelength (~~is depend on~~) **depends on** temperature.
- The line-edge roughness (~~should be depend on~~) **should depend on** the size of the polymer aggregates.

NOTE:「Depend on」の反意語は「**be independent of**」である。
- The characteristics should **be independent of** ambient conditions.

contain vs. include

POINT: あるものは他のものを含めている場合，「contain」は一番単純で，基本的なイメージを表す。一方，「include」を使うと，特別なニュアンスを持つ。

CONTAIN

DEFINITION: X contains A とは，AはXに入っていることを指す。

「X contains A」という表現を用いる場合，X に A だけが含まれているかどうかについては気にしていない。上図のように，「X contains A」に加えて，「X contains A and B」および「X contains A, B, and C」という表現も正しい。

- Type-A film **contains** more water than Type-B film.
  AタイプのフィルムはBタイプのフィルムに比べ水を多く含む。
- Each device **contains** a single nanometer-sized island.
  各デバイスにナノサイズの「島」が一つだけ存在する。
- The batteries should **not contain** any toxic substances.
  電池には有毒物質を含んではいけない。
- Fingerprint images usually **contain** stray black and white dots.
  指紋イメージには通常不規則な白黒の点が含まれる。

INCLUDE (1)

DEFINITION: まずあるグループに言及した後，それに含まれる一部（たまにすべて）のものを「include」を使ってリストアップすることができる。

グループ　　　　　　　　　　　　　リスト

- There are several large cities in Japan. They **include** Tokyo, Osaka, and Fukuoka.
- The next topic is patterning characteristics. They **include** resolution, mask linearity, critical-dimension control, and exposure latitude.
- The materials used to make optical waveguides **include** semiconductors, organics, and glasses.

INCLUDE (2)

DEFINITION: X includes B とは、他のもの（特に説明の必要はない）に加えて、B も X に入っていることを指す。

- The test setup **included** a newly developed circuit.
  試験装置には（標準的な部分に加えて、）新しく開発した回路も入っていた。
- The module **includes** a temperature control device.
- This unit **includes** an AC/DC converter.
  このユニットの標準的構成要素に加えて、AC/DCコンバーターもこのユニットに含まれている。

INCLUDE (3)

DEFINITION: A を X に include する、または A が X にincludeされるとは、意図的に A を X に入れることを意味する。

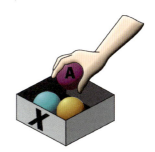

- For comparison, some previously reported data **are included** in the graph.
- There is also a polymer layer, but **I did not include** it in this diagram.
- For simplicity, no compensation circuits **were included**.

INCLUDE (4)

DEFINITION: 最初に全体量に言及する場合、ある量について「include」を用いると、それは全体の一部であり、全体とは別物ではないとの意味である。

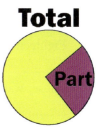

- The programs contain about 300,000 lines of code, **including** the 100,000 lines that we provided.
  全部で300 000行のプログラムのうち、われわれは100 000行を作成し、他の人は残りの200 000行を作成した。
- The recovery time **includes** a control time of 50 nanoseconds.

Section 2

> **PRACTICE:** 「Contain」か「include」、またはその両方を用いて空欄を埋めよ。両方を用いる場合、その意味の違いに気をつけよ。

1. Since this box _____ only two books, it is very light.
2. The development took 9 months, _____*ing* testing.
3. I haven't _____ them in the diagram to keep things simple.
4. This large structure in the crystal _____ an extremely large void (= empty space).
5. To check the fabrication process, small test elements are _____ on a chip.
6. These devices all _____ 50 transistors.
7. Next, some results for a fabricated device are presented. These _____ the conversion gain, intermodulation distortion ratio, and noise figure.
8. The switching time was estimated to be 60 ps, which _____ the 50-ps rise time of the electrical pulse generator. (ps = picoseconds)
9. These equations _____ very complex operations.
10. The buried oxide is 84 nm thick, _____*ing* a 20-nm-thick internal thermal-oxide (ITOX) layer.
11. Possible applications _____ satellite broadcasting and cable TV.
12. This structure has several advantages, _____*ing* reduced parasitic capacitances and simple device isolation.
13. The chips _____ 320,000 transistors.
14. A transistor _____ both n- and p-type regions.
15. Eighteen masks are used for the lithography. This number does not _____ the ones for wiring.
16. These wafers went through our LSI fabrication process, which _____ plasma etching, implantation doping, etc.
17. The drain saturation current of an nMOSFET _____ the parasitic bipolar current.

on the contrary

MISTAKE（間違い）	CORRECTION（訂正）
Japanese are poor at English.	 **On the contrary**, many Japanese speak English very well these days.

DEFINITION: いま述べられたことについてまったく賛成できないか**反対する**場合,「**on the contrary**」を用いる。[これとは反対に, それどころか]

「On the contrary」は, 直前に述べられたことに比べ, 単に何かが違うことをいう場合には使わない。直前に述べられたことが**間違い**だと主張するために使う。

技術英語では, この表現はほとんど使わない。

Good Examples

- **A:** Everyone can certainly afford soap to take a bath.
  B: On the contrary, a great number of people in India and elsewhere cannot even buy enough food!
- Of course, I'm not saying that this type of transistor is useless. **On the contrary**, I think it is one of the most promising designs.

~~on the contrary~~ → **in contrast (to)** / **by contrast (to)**

DEFINITION: 次に述べることが, 直前に述べたことと**大きく違う**場合,「**in contrast (to)**」または「**by contrast (to)**」という表現を用いる。[それと対照的に, それとは異なり]

Typical Mistakes

- For Process C, the <u>current decreases</u> between the first and second measurements. (~~On the contrary~~) **In contrast**, there is <u>no change</u> for Process D.
- The surface of the cathode was very <u>smooth</u>, and there was no apparent damage. (~~On the contrary~~) **In contrast,** the surface of the anode was <u>rough</u>.

Good Examples

- Many Type-A devices failed in less than 100 hours. **In contrast**, only one Type-B device failed in 750 hours.
- The wafer without the spacer exhibits a bumpy surface morphology. **In contrast**, the wafer with the spacer has a smooth surface.
- It is difficult to couple a waveguide-type avalanche photodiode to an optical fiber. **In contrast**, coupling is easy with the vertically illuminated type.
- The capacitor keeps the supply voltage under –45 V. **In contrast**, without a capacitor the supply voltage exceeds –30 V.

adopt

DEFINITION: ある問題への取り組み方，策略，計画，方法，方針などをadoptするとは，それを選択するまたは使うことにして実際に使い始めることである。

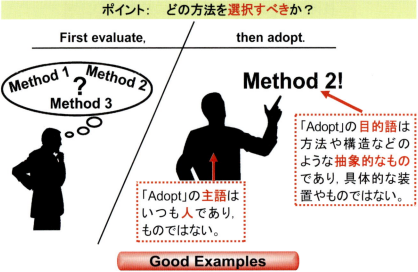

Good Examples

- To achieve these goals, we adopted a new architecture.
- The Standards Committee adopted our framework.
- One solution is to adopt a mixed design.
- The International Space Station has abandoned Microsoft Windows and has instead adopted LINUX.

ポイント： 実際に行ったことを説明する

実際に行ったこと(実験，計算，製作，設計など)を説明するとき，「**use**」または「**employ**」を使おう。

 ➡

Method 2 was **used** to …
Device A was **employed** to …

Typical Mistakes

✗ *These <u>machines</u> generally <u>adopt</u> superconducting bending magnets.*
「Adopt」の主語は人であるべき。
○ These machines generally **employ** superconducting bending magnets.

✗ *This illustrates the design method we <u>adopted</u>.*
何を行ったかを説明するだけであり，「adopt」はこのように使わない。
○ This illustrates the design method we **used**.

✗ *We <u>adopted</u> super-dynamic D-FFs because …*
物理的な装置はadoptできない。
○ We **employed/selected** super-dynamic D-FFs because …

✗ *To obtain high-speed operation, we <u>adopted</u> a gate width ratio of 0.5.*
数値はadoptできない。
○ … we **used** a gate width ratio of 0.5. / … we **set** the gate width ratio **to** 0.5.

✗ *This <u>amplifier</u> <u>adopts</u> a successive-detection architecture.*
「Adopt」の主語は人間であるべき。
○ This amplifier **employs/is based on** a successive-detection architecture.

cannot　can not　can't

POINT 1: 「Cannot」も「can not」も正しいが，「**cannot**」の方がよく使われているので，それは標準的な綴りとして覚えよう。

POINT 2: 技術英語で，「can't」，「don't」，「it'll」などの**短縮形**を使うのは避けた方がよい。

in case of fire

POINT:「In case」または「in case of」は異常が生じる場合だけに使われる。日本語の「場合」は「for」または「when」に訳そう。

DEFINITION:「In case of」は,特別な状況下で何をすべきかを示すものであり,公式な掲示に用いられている。[…の場合には,の際は]

- **In case of emergency**, call 110.

DEFINITION: 何か異常が起こる恐れのある場合,「in case」を使ってどうすれば安全かを示す。[もし…の場合には,もしも(万一)…なら,…の場合に備えて,…するといけないから]

- Take a map **in case** you get lost.
- Here is my telephone number **in case** you need to call me.

DEFINITION: 通常のケースと違う個別な処理方法などを,「in the case of」を使って示す。[…に関していえば,…については]

- It takes 4 years to graduate from university, or 5 **in the case of** premed students. (premed student = メディカル・スクール進学課程学生)
- Everyone who gets on the bus must pay the fare, **except in the case of** a woman carrying a baby, where the baby is free.

技術英語では,この三つの表現はほとんど使わない。

「In case」,「in case of」または「in the case of」の代わりに,「for」または「when」を使って条件を示そう。

✗ *In case of smaller circuits, SIMOX devices are 25% to 50% faster than bulk ones.*
「もし万一回路がもっと小さいならば」という意味となってしまう。
○ **For** smaller circuits, SIMOX devices are 25% to 50% faster than bulk ones.

- This is the calculated waveform (~~in the case of~~) **for** ideal boundary conditions.
- Figure 5 shows results (~~in case of~~) **for** a data rate of 1 Gb/s.

in case of fire   43

- WSiN makes a good barrier layer (*in case*) **when** an RF bias is applied during formation.
- (*In case*) **When** a positive bias is applied to the electrode, ions are deposited only at defect sites.

> **PRACTICE:** 下記の例にならい「for」または「when」を使って括弧内の条件を文に追加せよ。また，必要に応じて，=, <, >などの数学記号も英語の表現に直せ。
> Ex. 1: The flow rates were 6 sccm (SF$_6$) and 4 sccm (CF$_4$).
> ⇨ The flow rates were 6 sccm **for SF$_6$** and 4 sccm **for CF$_4$**.
> Ex. 2: (Si layer, thick), we get simple periodic oscillations.
> ⇨ **When the Si layer is thick**, we get simple periodic oscillations.

1. The values are 8.4 mW (Device A) and 11.2 mW (Device B).

2. This phenomenon appears only (the input power > a certain value).

3. This figure shows pulse patterns (output voltage = 200 mV)

4. (the voltage, low), light passes through.

5. These figures show the performance (156-Mb/s signals).

6. The resistivity increased monotonically with annealing time (zirconium layer, 200 nm thick).

7. (total thickness > 1 μm), the coupling efficiency increases dramatically.

8. Chemical polishing reduces the threshold voltage (long-channel devices).

9. (a defect is completely within the silicon), it does not give rise to a defect in the gate oxide.

10. (a conventional DFB laser), the lasing mode is very stable.

Connecting Nouns

POINT 1: 次のように名詞の前に名詞をおくことがある。
- ○ stone wall
- ○ Adidas leather tennis shoes

この場合，前方の名詞は最後の名詞の種類，特性などを示す。
しかし，次の表現は正しくない。
- ✗ *the room wall*

上の例では，room は wall の物質，目的，使用法，または他の特性を示さないので，通常**前置詞**を用いてそのような関係を示す。
- ○ the wall **of** the room

POINT 2: 多くの名詞を連ねた表現は通常わかりにくいので，避けた方がよい。
- ✗ *interlayer dielectric water molecule behavior*
- ○ the behavior **of** water molecules **in** interlayer dielectrics

以下の場合，「of」または「in」を使おう。

Components, parts, etc. (構成要素，部分など) (of)

the inner **wall of** the hole
the **top of** the valence band
each **stage of** the amplifier

the interface **circuits of** LSIs
the **surface of** the resist

Number, amount, size, etc. (数，量，サイズなど) (of)

the **number of** circuits
the **amount of** impact ionization
the increasing **size of** wafers

the **degree of** copper contamination
the **percentage of** people with allergies

Changes (変化) (in, of)

NOTE: an increase **in** voltage **of** 5 V
[Section 1の「**Changes & Differences**」の題を参照]

Properties (特性) (of)

the **thickness of** the layer
the **resistivity of** the wire

the **slope of** the line
the **amplitude of** the vibrations

Connecting Nouns 45

✘ property in something

水分 (moisture) は土 (soil) の中にある。しかし，水分の含有量 (moisture content) は具体的な物ではなく，数値であるので，土の中に存在するわけではない。それは土壌の特性 (property of soil) である。

- The moisture **content** (~~in~~) **of** the soil is 10%.
- The gain is related to the **carrier density** (~~in~~) **of** the active section.
  反応部分の中にあるのは電荷キャリアであり，その濃度ではない。
- The **refractive index** (~~in~~) **of** the waveguide decreases as the current increases.

Procedures, processes, etc. (手順，過程など) (of)

the **calculation of** very small values	**estimation of** the base resistance
the **regeneration of** optical signals	**measurement of** the optical properties
the **etching of** contact holes	**control of** the exposure conditions

Mathematical operations (数学演算) (of)

the **sum of** the harmonic components	the **ratio of** X to Y
the **product of** n^2 and the length	

NOTE: 各科学技術分野においてよく用いられる単語の組合せは，短縮され固有名詞として使用されるため，その標準的な組合せはその分野の慣習に従い使うべきである。このような表現は通常短く，二語の組合せになっている。関連分野の論文などを調べ，そのような表現を見つけて覚えておこう。

gate length	hole concentration
line width or linewidth	substrate temperature
crystal quality	signal-to-noise ratio, S/N ratio

PRACTICE: 例にならって，次の文に適切な前置詞を入れよ。

Examples:
a <u>new</u> contact hole etching <u>method</u> ⇨ a <u>new method</u> **of** etching contact holes

the oxide quality degradation ⇨ the <u>degradation</u> **in** the quality **of** the oxide

the field oxide edge ⇨ the <u>edge</u> **of** the field oxide

1. the Si-Si bond <u>length</u>

Section 2

2. the critical-dimension <u>change</u>

3. the intensity profile <u>calculation</u>

4. the mirror <u>scanning frequency</u>

5. the GaAs buffer layer <u>thickness</u>

6. sub-100-nm space pattern <u>replication</u>

7. the data traffic amount <u>increase</u>

8. at less-than-10-Mb/s <u>coding rates</u>

9. a surface-acid amount control <u>method</u>

apply

POINT: 次のページに示している「apply」のその他の使用法」以外には，「**apply**」の目的語はデバイス，IC，設備などのような具体的なもの（有形物）ではない。

DEFINITION: 仕事または活動にアイデア，プロセス，技術などを**apply**するとは，それを特定の仕事または活動に使用することである。

apply ＋ 抽象的なもの ＋ to ＋ 活動

Good Examples

apply idea   to fabrication

- To **apply** this <u>idea</u> **to** the <u>fabrication</u> of transistors, we need to use a low growth temperature.

apply lithography　　　　to production
- In order to **apply** X-ray lithography **to** the mass production of electronic devices, high-performance exposure characteristics are required.

apply techniques　　　　to design
- This section explains how to **apply** the techniques described above **to** the design of baseband circuits.

apply technique　　　　to assembly
- This packaging technique **has been applied to** the assembly of an all-optical wavelength converter module.

- ✗ *This amplifier can be applied to single-conversion transceivers.*
  有形物をapplyすることはできない。
- ○ This amplifier **can be used in** single-conversion transceivers.

- This transistor (~~can be applied to~~) **can be used to make** high-speed digital ICs.

- We have developed PR mapping equipment and (~~applied it to~~) **used it on** HEMT wafers.

「apply」のその他の使用法

◇ ルール，方式，数学的手順などを実際に当てはめる
- One approach is to apply a longest-path algorithm to the entire layout.
- First, a smoothing filter is applied to the captured image to eliminate stray white and black dots.

◇ …に電圧，力，圧力を印加する
- The ratio increases substantially when an RF bias is applied.
- Wafer distortion is caused by pressure applied to a limited area of a wafer.

◇ 液体などを物体の表面に塗る
- …apply a coat of resist to a wafer…
- Next, we apply an anti-reflection coating to the back of the substrate.

◇ 機械などを稼動する（注意：この使用法は技術英語ではほとんど使わない。また，この場合前置詞「to」は使わないこと）。
- To stop the car, just apply the brakes. （ブレーキをかける）

Hyphen (-)

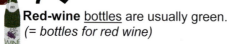

Red wine **bottles** are red.
(= red bottles for wine)

Red-wine <u>bottles</u> are usually green.
(= bottles for red wine)

> ハイフンがない場合，形容詞は最後の名詞を修飾する。

> **名詞の前**の複合形容詞の中の単語どうしの関連を示す場合，ハイフンを用いる。

NOTE: 名詞の前に置く名詞は単数形である。（✗ red-wine**s** bottles）

1. 数―名詞　／　数―名詞―形容詞

<u>3</u> computer **systems**

a **3-computer** system
(✗ a 3-computer**s** system)

> ハイフンがない場合，「3」は最後の名詞を修飾する。

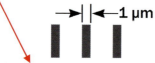

<u>3</u> µm-wide **lines**
(= three 1-µm-wide lines)

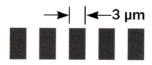

3-µm-wide lines

a 3-bedroom apartment 2-input NAND gates	a 3-region model a 50,000-word dictionary	a two-step process six-stage amplifiers
a 6-year-old child 0.3-µm L&S patterns	40-GHz optical signals 6-inch wafers	10-Gb/s modules 3-ps pulses

2. 形容詞―名詞　／　名詞―形容詞

high-quality wafers the soft-error rate high-energy photons	short-gate-length FETs real-time processing a bird's-eye view	a high-temperature process low-power, high-speed VLSIs
cross-sectional view error-free operation	a mirror-polished surface a mode-locked laser	hot-carrier-induced degradation surface-illuminated photodiodes

Punctuation: Hyphen (-) 49

3. 形容詞として使われている句

electron-hole pair	time-space conversion	current-voltage
via-to-via pitch	state-of-the-art technology	characteristics
on-wafer probe	on-demand manufacturing	analog-to-digital converter

4. 句動詞から作られた名詞または形容詞

a trade-off	lift-off	a plug-in	turn-around time
a blow-up	hole pile-up	an add-on	turn-on voltage

「Setup」、「breakdown」などよく使われている名詞はハイフンがなくなった。正しい綴りを辞書でチェックする。

> **NOTE:** ハイフンを含む表現をリストアップする場合、ハイフンをそのまま残す必要がある。
> zero-dimensional and two-dimensional structures
> ⇨ zero- and two-dimensional structures
>
> 3-μm-wide, 5-μm-wide, and 7-μm-wide lines
> ⇨ 3-, 5-, and 7-μm-wide lines

5. NO HYPHEN

1. 名詞が並んだ場合、通常最後の名詞の前にはハイフンをつけないこと。
   post office   train station   wire bonding   quantum well   InP wafer

2. 接頭辞の後ろには一般的にハイフンをつけないこと。
   pseudorandom   multichannel   optoelectronic   postprocessing
   特にSI単位系接頭辞（*centi-, milli-, mega-, giga-* など）の後ろにハイフンをつけないこと。
   gigabit   kilovolt   milligram   picofarad   nanometer   (✗ *nano-meter*)

3. 例外：発音をはっきりさせるために接頭辞の後ろにハイフンをつけることがある。
   electro-optical   multi-electrode   pre-emptive

PRACTICE: 次の文には誤ったものが含まれている。それを見つけて直せ。

1. We made 3-μm long devices.
2. We made 3-μm-long devices.
3. We made 3-microns-long devices.
4. We made a device 3 μm long.
5. We made a device 3-μm-long.
6. a thick buffer layer
7. a 15-minutes presentation
8. a 1-V power supply
9. a low voltage LSI
10. a low voltage level
11. a gate-width of 5 μm
12. at a low-V_{th}
13. a silicon-on-insulator substrate
14. a high temperature process
15. a one dimensional system
16. a GaAs buffer layer 480-nm thick
17. a 0.5 μm thick layer of resist
18. lattice matched substrates
19. Each device is 2.5 mm-long.
20. a high operating frequency of 200-MHz

50   Section 2

Dynamic Verbs 2

能動態は受動態より説得力がある。

次の例を考えよう。
- The electrical characteristics are not changed by adding the intermediate layers.
- The intermediate layers do not change the electrical characteristics.

2番目の文は簡潔で明白であり，また直接的である。それは能動態のパワーを示している。

PRACTICE: 下線の引いてある動詞を能動態に変えよ。

1. In the calculations, these atoms <u>are taken</u> into account.

   The calculations _____

2. Mechanical stress <u>is generated by</u> wire bonding.

   Wire bonding _____

3. The light <u>was focused</u> on the device with a lens.

   A lens _____

4. Voltage oscillations <u>are suppressed by</u> the load capacitance.

   The load capacitance _____

5. The characteristics <u>are</u> dramatically <u>improved by</u> the feedback.

   The feedback _____

6. The temperature <u>is raised by</u> the light.

   The light _____

7. In this electrical excitation, ballistic electrons <u>are generated</u>.

   This electrical excitation _____

8. In this model, the gate-drain capacitance <u>is included</u>.

   This model _____

9. With this circuit, the number of connections <u>is reduced by</u> 75%.

   This circuit _____

10. In Section 4, the effectiveness of the method is demonstrated by numerical examples.

    In Section 4, numerical examples _____

Prepositions 2

空欄を適切な前置詞で埋めよ。必要がなければ，×で埋めよ。

REVIEW

a. X is due _____ Y.

b. X influences _____ Y.

c. X equals _____ Y.

d. This paper discusses _____ a new approach...

e. X is the same _____ Y.

CHECK YOUR KNOWLEDGE

1. X is independent _____ Y.

2. X has an influence _____ Y.

3. X causes _____ Y.

4. X is identical _____ Y.

5. X is equivalent _____ Y.

6. Result X agrees _____ result Y.

7. X affects _____ Y.

8. X has an effect _____ Y.

9. X is suitable _____ Y.

10. X is composed _____ Y and Z.

Section 3

compared to vs. than

damage vs. damages

for −ing

is expected

approach

can   could

consist of

as a result

proportion(al)

is thought

each

prepare

becomes vs. is

Punctuation: Colon (:)

Style: Unnecessary Repetition

Prepositions 3

compared to vs. than

比較する場合，通常比較形(-er than)を用いる。

- ✗ *X has a higher density compared to Y.*
- ○ X has a **higher density than** Y.   X is **denser than** Y.

- ✗ *Compared to millimeter waves, terahertz waves are shorter.*
- ○ Terahertz waves **are shorter than** millimeter waves.

- ✗ *The bandwidth of X is improved by 75% compared to that of Y.*
- ○ The bandwidth of X is **75% larger than** that of Y.

比較の基準を示す場合，「compared to/with」を用いる。

「犬は大きいですか，それとも小さいですか？」
　この質問は，比較の基準を示さない限り，無意味である。

A dog is **big**, **compared to** a mouse.

NOTE: 「Bigger」ではなく，「big」である。

NOTE: 「Smaller」ではなく，「small」である。

A dog is **small**, **compared to** an elephant.

Good Examples

- The grains in the crystal are relatively **small, compared to** those of the reference sample.
- These devices exhibit very **fast** operation, **compared to** silica-based ones.
- The radius is very **small**, **compared to** the length.
- Freezing is a **relatively minor** factor, **compared to** moisture migration.
- The diffusion length of Zn is **negligible, compared with** the size of the device.
- **Compared with** xylene, ester acetates have a rather **low** sensitivity; but they provide better contrast.

Section 3

> **PRACTICE:**「Compared」または「comparison」を削除せよ。

> **NOTE:** 形容詞を強調するとき，「very big」のように「**very**」をよく用いる。比較形の場合，強調するとき，「much bigger」または「much more beautiful」のように「**much**」を用いる。

1. This method is **very effective** <u>compared to</u> conventional ones.

2. The processing power is **very low** <u>in comparison with</u> that of conventional machines.

3. Holes have a **large** effective mass <u>compared to</u> electrons.

4. The gains are **very high** <u>compared to</u> this value.

5. The size of the new circuit **is reduced by about 25%** <u>compared with</u> that of a conventional one.

6. The accuracy is **lower** <u>compared to</u> that of the simulation results.

damage vs. damages

> **POINT:** 技術論文では，「damage」は常に単数形である。複数形の「**damages**」はお金を意味する。

> **DEFINITION:**
> 1. 「**Damage**」は物に与えられた物理的損害・損傷である。
> 2. 例えばAさんがBさんに損害や傷害を加えたとする。裁判所命令によってAさんがBさんに払わなければならない損害賠償金は「**damages**」である。

- Voltage drops and other disturbances in a utility grid may cause severe (~~damages~~) **damage** to sensitive local loads.
- This paper concerns a new dry-etching process without any (~~damages~~) **damage**.
- (~~Damages~~) **Damage** to telecom equipment due to weather is a serious problem.

for –ing

POINT: I went to the store **for buying bread**.
この文は正しい英語ではない。

動作の目的を示す場合，右のような表現を用いる。

to + verb (目的を示す不定詞)
in order to + verb
so that, etc.

○ I went to the store
- **to buy** bread.
- **in order to buy** bread.
- **so that I could buy** bread.

✗ *For fabricating the devices, we selected UV-curable resin* (紫外線硬化樹脂).
以下に示されているような熟語の一部ではない限り，「for -ing」で文を始めるのはよく見られる間違いである。
○ **To fabricate** the devices, we **selected** UV-curable resin.
- Two technologies **were developed** (~~for achieving~~) **to achieve** fast access to external memory.
- The temperature of the laser **is changed** (~~for tuning~~) **to tune** the wavelength.

物の用途を示す場合，「for -ing」を用いることができる。

- This is a **lens for focusing** light.
- There are two conventional **devices for connecting** optical fibers.
- Figure 6 shows the **setup for testing** the modulator.
- Several **formulas for calculating** maximum heart rate have been reported.
- The gold film provides a **surface for sensing** the refractive index.

熟語の中の「for -ing」はこの限りではない。

use for, be used for
- This system **is used for delivering** high-definition TV programs.
- The structure **used for sealing** can affect the transmission characteristics
- **For imaging** concealed cracks in concrete, a millimeter-wave antenna **is used**.
熟語の一部であれば，「for –ing」で文を始めることができる。

be 形容詞 for (形容詞 = good, useful, suitable, necessary, etc.)
- Si wire waveguides **are good for making** compact devices.
- Terahertz spectroscopy **is useful for studying** the vibrational modes of weak chemical bonds
- This circuit **is suitable for estimating** the waveform of capacitance oscillations.
- The functions **necessary for generating** the signals are integrated on a small chip.

need for
- The injection current **needed for switching** is very low.
- **For sensing**, we **need** a wide frequency tuning range and fast frequency sweeping.

thank for
- The authors wish to **thank** K. Suzuki **for making** the X-ray masks.

以上のような熟語はたくさんあることに注意しよう。

PRACTICE: 次の文中で間違っているのはどれか？それらを正しい文に直せ。

1. In this LSI, four metal layers are used <u>for wiring</u>.
2. The layer structure was designed <u>for achieving</u> a fast response.
3. This section describes pulse-pattern generators <u>for testing</u> ultrafast ICs.
4. A thermoelectric cooler was added <u>for controlling</u> the temperature of the chip.
5. <u>For tracking</u> moving objects, it is necessary to match key features in neighboring frames. (frame = コマ)
6. The injection current needed <u>for switching</u> is very small.
7. Figure 3 shows a frame <u>for supporting</u> 16 fibers. (frame = 枠)
8. Experiments were performed <u>for verifying</u> the proper operation of the device.
9. We optimized the laser <u>for improving</u> the performance of the laser array.
10. <u>For measuring</u> the resistance, the cathode was attached to a Pt mesh.
11. We used a laser pointer as an ideal device <u>for pointing</u>.
12. Atomic-force microscopy (原子間力顕微鏡法) is now widely used <u>for examining</u> surface structures.
13. The composition of the WSiN was measured <u>for determining</u> the effect of an RF bias.
14. The criterion <u>for determining</u> the pedal load is based on heart rate.

is expected

POINT: 著者が個人的に expect しただけなら，この表現は使わず，「**should**」を用いる。

DEFINITION: 何かが **is expected** ならば，多くの人はそれが起こるのを予想できている。すなわち，その発生する可能性があることは一般的に認識されている。

技術英語では，「be expected」はあまり使わない。

Good Examples

- The trend toward further device miniaturization **is expected to continue**.
  だれもがトランジスタが小さくなることを信じている。
- The market for ultrahigh-definition TV **is expected to grow** rapidly.
- The number and variety of mobile devices **are expected to increase** greatly.
- Ultralow-voltage circuit technology **is expected to pave the way** to mobile solar-cell systems.

Typical Mistakes

- ✗ *This device is expected to operate at high bit rates of over 100 Gb/s.*
  このデバイスを作成した研究者本人だけがそう思っているため，一般的に認められている考えではない。
- ○ This device **should operate** at high bit rates of over 100 Gb/s.

- Examining this region (~~is expected to give~~) **should give** us information on the behavior of the ethoxy group (C_2H_5O).

- This technique (~~is expected to fill~~) **should fill** even deep holes and gaps.
- When these molecules aggregate, (~~it is expected that they~~) **they should** become smaller.

期待できる ➡ promising

- ✗ *The resonant tunneling diode <u>is expected as</u> the most useful quantum effect device.*
- ○ The resonant tunneling diode is the most **promising** quantum effect device.

NOTE: ここで問題になっているのは「be expected」であり、以下の「expected」の使い方には特に問題はない。

- The intensity is much **lower than expected**.
- These data show the resolution **expected from** Eq. 1.
- This value is **smaller than that expected from** the simulation.
- If the slope errors were comparable to this value, **the expected** exposure characteristics would not be obtained at all.
- **As expected**, the ratio slowly drops as time passes.

approach

名詞

POINT: 「Approach」の後ろの「**to**」は前置詞であるため、その後ろには、動詞ではなく、名詞または動名詞を使わなければならない。

- ○ an approach to the problem　　　　　to + 名詞
- ○ an approach to solving the problem　　to + 動名詞
- ✗ an approach ~~to solve~~ the problem　　不定詞
- ✗ an approach ~~for solving~~ the problem　for + ___ing

- A new **approach** (~~to remove~~) **to removing** the polymer layer is needed.
- One **approach** (~~to expand~~) **to expanding** network capacity is Si photonics technology.
- This is a common **approach** (~~to make~~) **to making** such systems robust against failure.

動詞

POINT: 「Approach」は他動詞であるので、その後ろには前置詞の「to」ではなく、目的語を使わなければならない。

- ✗ X approaches ~~to~~ Y.
- The robot **approaches** (~~to~~) the straight-up equilibrium position.
- $\cos\theta$ **approaches** (~~to~~) one as θ tends toward zero.

can　could

Ex. 1: まず，次の文を考えよ。
- Bob **can run** the 100-meter dash in 10 seconds.（ボブは100メートル走を10秒で走ることができる。）

この文について，次のような疑問が生じる。
- 本当？ボブは実際に100メートル走をしたことがあるか？
- なぜ10秒で走れることがわかるか？
- それは事実である証拠はあるか，それとも取るに足りない自慢話だけか？

次に，以下の「事実を述べる文」を考えよ。
- Bob **ran** the 100-meter dash in 10 seconds.（ボブは100メートル走を10秒で走った。）

この記述では，次のことについて疑う余地はない。
- ボブはそれをすることは可能である。
- ボブはそれをする能力がある。

Ex. 2: 次の文の違いを吟味せよ。
- The sun **can rise** in the east.　　（太陽は東から昇る可能性がある。）
- The sun **rises** in the east.　　　（太陽は東から昇る。）

起こる可能性があるということは，起こらない可能性もあるということである。そのため，実際に起こっていることについては「can」は使えない。

一般的に，研究結果，発見，観察および結論を「can」か「could」を用いて述べるよりは，事実として述べた方がはるかに説得力がある。

Typical Mistakes

次の表現に特に注意を払おう。

 ➡
can/could achieve　　　　achieved

✗ *We <u>can achieve</u> a high coupling efficiency of over 70%.*
　可能性？どの条件のもとで？実際にやったことあるか？
○ We **achieved** a high coupling efficiency of over 70%. (Fact!)

✗ *We <u>could achieve</u> a large tolerance of about 50 μm.*
○ We **achieved** a large tolerance of about 50 μm. (Fact!)

Section 3

 ~~can/could obtain~~ ➔ **obtained**

- ✗ *For the quarter-micron SOI process, we <u>can obtain</u> a delay time of 10 ns at a voltage of 0.5 V.* （可能性？）
- ○ For the quarter-micron SOI process, we **obtained** a delay time of 10 ns at a voltage of 0.5 V. (Fact!)

- ✗ *By optimizing the O_2 flow rate, we <u>could obtain</u> high-quality films.*
- ○ By optimizing the O_2 flow rate, we **obtained** high-quality films. (Fact!)

- ✗ *A dynamic range of over 28 dB <u>could be obtained</u> for the module.*
- ○ A dynamic range of over 28 dB **was obtained** for the module. (Fact!)

 ~~can/could observe~~ ➔ **observed**

- ✗ *We <u>could observe</u> a melted region on the input facet.*
- ○ We **observed** a melted region on the input facet. (Fact!)

他の例

- ✗ *We have recently developed a DBR laser that <u>can provide</u> a mode-hop-free tuning range of more than 6 nm.*
  そうなることも，ならないこともあるのか？どの条件のもとで？
- ○ We have recently developed a DBR laser that **provides** a mode-hop-free tuning range of more than 6 nm. (Fact!)

- ✗ *This HEMT <u>can satisfy</u> the requirements for 100-Gb/s ICs.*
  そうでない可能性もある！
- ○ This HEMT **satisfies** the requirements for 100-Gb/s ICs. (Fact!)

- Each microlens (~~could be formed~~) **was formed** at the correct position.
- The circuit (~~could generate~~) **generated** an ultrafast electrical pulse.
- We (~~could demonstrate~~) **demonstrated** the CW operation of the LDs at room temperature.
- The median lifetime (~~could be calculated~~) **was calculated** from the activation energy.

Good Examples

- This device **can produce** a conversion gain when an intense pump beam is injected.
  [能力]（必要ならば，このデバイスは変換ゲインを出すことができる。）
- Millimeter-wave signals **can be generated** at frequencies of up to 240 GHz by using a mode-locked laser.
  [可能性]（可能性の限界を示す）

- We think we **can increase** the output power by optimizing the antenna design.
  [可能性]（出力のパワーを増やすことが可能であるが、それについて絶対的な確信は持っていない。）

> **PRACTICE:** 次の文中で間違っているのはどれか？それらを正しい文に直せ。

1. The good agreement between the experimental and simulation results **can demonstrate** that the simulation method is correct.
2. These switches **can operate** at bit rates as high as 20 Gb/s.
3. We **can use** either Method A or Method B to make the device.
4. The resolution **can be estimated to be** 0.19 mm from the dimensions of the device.
5. This fabrication technique **can prevent** the unintentional formation of parasitic islands.
6. WSiN **can prevent** copper diffusion.
7. **There can be** three possibilities: (a) no diode switches, (b) one diode switches, and (c) two diodes switch.
8. Each signal **can be input** either with or without a routing bit.
9. Silicon **can block** high-energy X-rays.
10. These results demonstrate that our algorithm **can improve** the verification accuracy.
11. This voltage region **can be divided** into three parts at the voltages V_1 and V_2.
12. We **can obtain** an output power of 1.7 W.

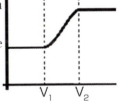

consist of

> **POINT:**「Consist of」と「be composed of」を混同してはいけない。

○ Z **consists of** X and Y.   ✗ ~~Z is consisted of~~ X and Y.
○ Z **is composed of** X and Y.   **NOT ENGLISH!**

- The battery pack (~~is consisted~~) **consists of** two parallel sets of 15 cells in series.
- The optical interconnections (~~was consisted~~) **consist of** SiO_2 and polymer.

as a result

POINT: 「**As a result**」は「**because**」の逆である。両方の表現は因果関係を説明するのに使われる。

as a result:
原因を最初に記述する。

 The vase fell. **As a result**, it broke.

because:
通常，結果を最初に記述する。

 The vase broke **because** it fell.

Good Examples

- The widespread use of Internet services <u>has produced explosive growth</u> in the amount of data transmitted. **As a result**, current telecommunications networks face a number of problems.
    - = Current telecommunications networks face a number of problems **because** the widespread use of Internet services has produced explosive growth in the amount of data transmitted.

- As the temperature increases, the substrate <u>expands laterally</u>. **As a result**, the waveguides shift relative to each other.
    - = As the temperature increases, the waveguides shift relative to each other **because** the substrate expands laterally.

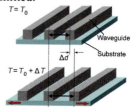

- This technique <u>requires only one measurement</u> of the system characteristics. **As a result**, the design and evaluation process is short.
    - = The design and evaluation process is short **because** this technique requires only one measurement of the system characteristics.

実験，テスト，計算などの結果を導入する場合，「**as a result**」は使わないこと。

Typical Mistakes

✗ *A test chip was evaluated. <u>As a result</u>, the new circuit consumed less power than a conventional one.*
新しい回路の電力消費量が少ないのは，測定されたことによるものではなく，設計の良さによるものである。

○ A test chip was evaluated. **The results showed that** the new circuit consumed less power than a conventional one.

○ An evaluation of a test chip **revealed that** the new circuit consumed less power than a conventional one.

- ✗ *I-V measurements were made on mesa structures 200 μm in diameter. <u>As a result</u>, high resistivities above 5×10^7 Ωcm were obtained.*
- ○ I-V measurements were made on mesa structures 200 μm in diameter, **and** high resistivities above 5×10^7 Ωcm **were obtained**.

> **PRACTICE:** 次の文において，正しいものを選べ。

1. The actuation voltage was swept up from 0 to 20 V and then down to 0 V. <u>As a result</u>, the switch turned on at a voltage of 10 V and turned off at 9 V.
2. This device reduced the power of the noise but not the power of the signal. <u>As a result</u>, the signal-to-noise ratio increased by 18 dB.
3. <u>As a result</u> of BET measurements, the specific surface areas of the oxides have the values listed in Table 3.
4. The coating protrudes slightly beyond the end of the glass fiber; and <u>as a result</u>, there is a gap when two fibers are connected.
5. Section 3 describes a high-speed optical-coherence-tomography (OCT) system. <u>As a result</u>, Section 4 presents OCT images made with the system.
6. The computational complexity is very low. <u>As a result</u>, computations are very fast.
7. Figure 7 shows XRD patterns of $Mn_{2-x}Fe_xO_3$ heated to 500°C. <u>As a result</u>, when x was 0, 0.2, or 0.4, a solid solution of Mn_2O_3 formed.
8. New optical networks require more complex processing on a larger scale. <u>As a result</u>, the number of optical devices in equipment is increasing.
9. Charging easily decomposes these very fine particles. <u>As a result</u>, the charging voltage is low.

proportion(al)

in proportion to
前置詞 ＋ 名詞 ＋ to

✗ ~~in proportional to~~
前置詞 ＋ 形容詞 ＋ to

- The current of a macrocell <u>increases **in proportion to**</u> its bandwidth.
- The power <u>decreases **in proportion to**</u> the square of the supply voltage.

be proportional to
be ＋ 形容詞 ＋ to

同じ構造を持つ表現： **be equal to** / **be identical to**

- The CMOS power consumption **is proportional to** the clock frequency.
- The resistance of a via-plug **is inversely proportional to** the square of the diameter.

is thought

POINT: 「と思われる」をそのまま「is thought」と訳すと，ほとんどの場合間違ってしまう。

DEFINITION: 「Is thought」は，ある分野の人々の共通する考え方を指す。すなわち，その考え方は一般的に認められているものである。

自分の実験結果または個人の意見を述べるとき，「is thought」は使わない。

✗ *This slight difference <u>is thought</u> to be caused by the difference in the sizes of the one-dimensional wires.*
この記述は研究者の実験結果に関するものである。この結果が発表されるまで他の人は知らないため，それは一般的に認められたものではない。

NOTE: 技術日本語において「と思われる」という表現は，ある発案やアイデアが正しそうな気配が濃いのに，それを証明できないときによく使われている。英語では，このようなケースをほとんど「probably」を用いて表現できる。それが適切でなければ，「seem」または他の表現を用いればよい。

○ This slight difference is **probably** caused by the difference in…

○ This slight difference **seems** to be caused by the difference in…

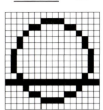

Typical Mistakes

- In the degraded image, one row is completely black. (*This is thought to be*) **This is probably** due to a short circuit (ショート).
- Since blister-like bubbles appeared after curing, this film (*is thought to contain*) **probably contains** a large amount of solvent.
- This film was deposited at a temperature of 150°C, which (*is thought to be*) **seems to be** the optimum deposition temperature.

Good Examples

新しいアイデアを一般的に認められている考え方と対照させる場合，まず，「is thought」を用いて一般的な考え方を提示し，次に，「**but**」あるいは「**however**」などで始まる文により一般的な考え方と違う新しいアイデアを示す。

- **Previously, it was thought that** annealing just caused the shrinkage and disappearance of small clusters to establish a local thermodynamic equilibrium. **But…**
- **It was thought that** making a practical high-voltage electron gun required many technical advances. **However,…**
- **Generally, it is thought that** development （現像）proceeds through the dissolution （溶解）of resist molecules decomposed by the e-beam. **However, our results do not support that idea. We found that…**

each

each + 単数名詞

- The peak of **each** (*spectra are*) **spectrum is** flat.
- **Each** (*PCs download*) **PC downloads** programs and data when necessary.
- **Each** (*devices have*) **device has** sufficient wavelength accuracy.

- (*The each*) **Each** arm of the modulator is driven by…

prepare

POINT:「Prepare」は通常何かを新たに製作するという意味ではない。

基本的に，ものをprepareするとは，未来に起こる出来事のために，そのものを**適した状態にする**ことである。そのものは**既に存在し**，新しく作られていないことに注意しよう。

- They **prepared** the room for the meeting.
  彼らは新しい部屋を造ったわけではない。その部屋は既に存在している。その人の行ったことは，椅子を並べ，カップと水差しを用意し，プロジェクターをセットしただけである。
- The equipment **was prepared** for the measurements.
  これらの機器は既存のものであり，測定に先立って整備，調整や較正などをしただけである。
- The samples **were prepared** for observation.
  サンプルは測定に先立ち，あらかじめ測定に必要なクリーニングや切断，研磨，化学処理などを行っただけである。

Typical Mistakes

prepare ≠ make fabricate create form

TEST: 「Prepare」の代わりに，「fabricate」，「make」，「create」，または「form」などの単語を用いると文が意味を持つなら，「prepare」は間違いである。

- Two types of transistors were (~~prepared~~) **fabricated**.
- An array of circular contacts was (~~prepared~~) **fabricated**.
- A combined instruction that executes two operations was (~~prepared~~) **created**.
- Thin films were (~~prepared~~) **formed** by spin coating a toluene solution of a polymer onto a substrate.
- A beat signal was (~~prepared~~) **generated** by combining two laser beams.
- A fiber crystal was (~~prepared~~) **grown** by the laser-assisted pedestal growth method.

例外：以下の三つの例で，新しいものが創りだされている。

1. 溶液，化合物，薬などを prepare するには，様々な物質を混ぜ合わせたり処理したりする。
   - Two kinds of tablets **were prepared.**

2. スピーチやプレゼンテーションが prepare できる。
3. 学校の先生が教材，宿題，および試験を prepare する。

becomes vs. is

POINT: 「To **become**」は常に変化を示し、be-動詞は状態を示す。

変化

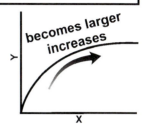

変化の表現:
- Y becomes larger **as** X becomes larger.
- Y increases **as** X increases.

「**As**」の使用法に注意。

点

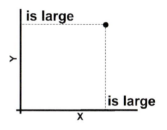

点の表現:
- Y is large **when** X is large.

NOTE: 「Becomes」と「is」を一緒に使用するのは通常好ましくない。

- ✗ Y *becomes* large when X *is* large.
- ✗ Y *is* large when/as X *increases*.

Typical Mistakes

- ✗ *For these applications, it becomes important to reduce the temperature sensitivity.* （今重要ではないが、いずれ重要になるという意味か？）
- ○ For these applications, it **is** important to reduce the temperature sensitivity. （今重要である。）

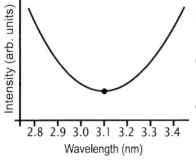

- ✗ *The intensity becomes a minimum when the wavelength is 3.1 nm.*
- ○ The intensity **is** a minimum **when** the wavelength **is** 3.1 nm.
- ○ The intensity **becomes** a minimum **as** the wavelength **approaches** 3.1 nm.

Section 3

- ✗ <u>When</u> the amount of data <u>is</u> large, the transfer time <u>increases</u>.
- ○ **As** the amount of data **becomes** larger, the transfer time **increases**.

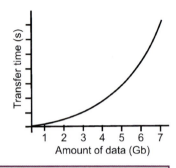

> **PRACTICE:** 次の文には誤ったものが含まれている。それを見つけて直せ。

PART A

1. The variation <u>becomes</u> worse as the voltage <u>increases</u>.
2. When V_{DD} <u>is</u> small, the delay <u>increases</u>.
3. If X <u>is</u> large enough, Z <u>becomes</u> negligible.
4. As X <u>becomes</u> stronger, Y <u>increases</u>.

PART B

1. This process <u>has become</u> the conventional way of cleaning silicon wafers because of its excellent performance.
2. With this design, the gain of one stage <u>becomes</u> G_{total}/n, where G_{total} is the total gain of the amplifier and n is the number of stages.
3. The poor reliability of tungsten plugs <u>has already become</u> a serious problem in multilevel interconnections.
4. To track moving objects, it <u>becomes</u> necessary to match feature points in neighboring frames.
5. Resists of this kind <u>are becoming</u> increasingly important in deep submicron lithography.

Punctuation: Colon (:)   69

Colon (:)

コロンの前にスペースを入れないこと。

1. リスト（列挙項目）の導入

- There are three main sources of power consumption in digital CMOS circuits: switching power, short-circuit power, and leakage power.

　　　　　[**Section 2** の「**Lists**」の題を参照]

2. 「As follows」の後ろには必ず使い，また「following」の後ろにもよく使う。

- The algorithm is **as follows**:
(1) Get error value, E.
(2) Rotate joint by +1° and render.
(3) Get new E.
- This graph can be understood **as follows**: The X-axis is the timing of the input data, and the Y-axis is the bit rate.
  コロンの後ろに完成文がある場合，その文は大文字で始まる。
- The stress (σ) is related to the height (H) and width (W) of a line by the **following** equation:
$$\sigma = \frac{6\gamma \cos\theta}{D}\left(\frac{H}{W}\right)^2 \tag{4}$$

3. 方程式の導入

方程式の前のテキストは完成文である場合，コロンを使用する。

- A full state observer with gain Ψ was used in this study:
$$\begin{cases} \hat{x}(t) = A\hat{x}(t) + Bu_f(t) + \Psi[y(t) - \hat{y}(t)] \\ \hat{y}(t) = C\hat{x}(t) \end{cases} \tag{17}$$

方程式の前のテキストは完成文でない場合，コロンを使用しない。

- Combining the full-state observer with (8) and solving for $\hat{d}_e(t)$ yield
$$\hat{d}_e(t) = B^+\Psi C[x(t) - \hat{x}(t)] + u_f(t) - u(t). \tag{18}$$

4. 数量または定義の導入

- The flow cell consists of a base plate, adhesive tape (thickness: 70 μm) with a flow channel pattern, and a lid.
- The robot arm is driven by a dc motor (rated voltage: 3 V; rated current: 0.56 A; rated speed: 895 rad/s).
- Table 1. Data on 47 subjects in experiment (SD: standard deviation).

Clear aperture:	12 mm²
Sensitivity:	100 mW/cm²

Clear aperture:	12 mm²
Sensitivity:	100 mW/cm²

コロンの前にスペースを入れないこと。

Unnecessary Repetition

代名詞を使って，同じ言葉の繰り返しを避けよう。場合によっては，文を再構築するように工夫しよう。

BAD: SiO_2 mask stripes are formed on a (100) n-InP substrate. The mask stripes are oriented parallel to the [011] direction. The spacing between the mask stripes is 2 μm.

GOOD: SiO_2 mask stripes are formed on a (100) n-InP substrate. **They** are oriented parallel to the [011] direction **and** have a spacing of 2 μm.

BETTER: SiO_2 mask stripes spaced 2 μm apart are formed on a (100) n-InP substrate parallel to the [011] direction.

PRACTICE: 不必要な繰り返しをなくせ。

1. We paid special attention to the influence of the subband system, and to the possibility of observing the subband system at high temperatures.

2. The overcoat was about 0.1 μm thick. The overcoat was removed after baking.

3. This mesa structure is not easy to bury because the mesa structure has no mask on top.

4. This is an example of an action logic table. This action logic table can also be transformed into Prolog.

5. We developed a global router for high-speed bipolar LSIs. This global router minimizes areas...

6. We observed the spots to estimate the amount of relaxation in the lattice parameters. The results of the observation of the spots revealed that the lattice parameters are relaxed in two steps.

7. This is the circuit we designed. The circuit was fabricated on a CMOS process.

Prepositions 3

空欄を適切な前置詞で埋めよ。必要がなければ、×で埋めよ。

REVIEW

a. X is equal _____ Y.

b. X has an effect _____ Y.

c. This section explains _____ the method.

d. Result X agrees _____ result Y.

e. X is suitable _____ Y.

f. X has an influence _____ Y.

g. X causes _____ Y.

h. X depends _____ Y.

i. X is called _____ Y.

j. X consists _____ Y and Z.

CHECK YOUR KNOWLEDGE

1. X is responsible _____ Y.

2. X is similar _____ Y.

3. This paper concerns _____ X.

4. _____ the other hand, ...

5. a change _____ the voltage

6. X is capable _____ Y.

7. X is in good agreement _____ Y.

8. X is different _____ Y.

9. X is _____ the order _____ 10^6.

10. X corresponds _____ Y.

Section 4

remarkable

control

tolerance

respectively

common vs. popular

recently

simplified

introduce

enough

Adjective Formation (-ing)

Adjective Formation (-ed)

compensate：他動詞 vs. 自動詞

conventional

_____

Punctuation: Comma 1

Style: Unnecessary Words 1

Prepositions 4

remarkable

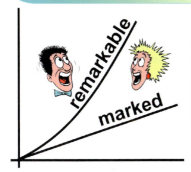

DEFINITION 1: Remarkable なこととは，例外的あるいは特別なことであり，人々を仰天させ，強く印象づけることである。[驚くべき，印象的な，異常な，例外的な，珍しい] **remarkably** (副詞)

「Remarkable」はときどきその分野において驚くべき進歩を記述するために論文の序論で使用される。それ以外に，**技術英語ではあまり使わない**。

- There has been **remarkable progress** in broadband communications.
- The **remarkable speed-up** of fiber-optic networks has spurred demand for correspondingly higher data rates in wireless transmission and access systems.

DEFINITION 2: Marked な変化または違いとは，非常に目立った明らかな変化または違いである。 **markedly** (副詞)

MARKED:

- There was **a marked increase** in the base current.
- Above the critical temperature, there is **a marked reduction** in the efficiency.
- We observed **a marked improvement** in the characteristics.
- There are **marked differences** between the two devices.
- The non-linearity **is more marked** for mid-size patterns.

MARKEDLY:

- As the growth temperature falls, the roughness **increases markedly**.
- This design **markedly reduces** spherical aberration.
- Annular illumination **does not markedly improve** the ultimate resolution.
- The transmission loss in smoke is **markedly smaller** for terahertz waves than for infrared light

NOTE:「Marked」の代わりに以下の言葉も使える。
    observable   noticeable   significant   pronounced

control

Ex. 1: 車のハンドル操作を考えよう。ハンドルで，車の進行方向を control する。「Control する」には二つの意味がある。

車を直進させる場合，ハンドルをそのまま固定し，すなわち，ハンドルを**切らない**。

車の方向を変える場合，ハンドルを**切る**。

Ex. 2: エアコンによる室温制御を考えよう。室温が高い場合，エアコンは温度を**変えて**部屋を涼しくする。室温が適切ならば，温度を**変えない**（維持する）。

したがって「**control**」とは望みのことをする**能力を発揮する**ことを指しているだけであり，何かが実際に変化することを意味するとは限らない。変化させたくないときにはそれを維持することができ，変化させたいときには変えることができる。

control ≠ change, adjust

Typical Mistakes

- ✗ *You make a car turn by <u>controlling</u> the steering wheel.*
- ○ You make a car turn by **turning** the steering wheel.
- ○ You **control the direction** in which a car moves by means of the steering wheel.

- ✗ *The lasing wavelength can be changed very rapidly by <u>controlling</u> the currents to the three electrodes.*
  電流を control するとはそれを変える，または一定値に維持する能力を発揮することであり，それが実際に変化している意味ではないため，この文は意味を持たない。

tolerance

- ○ The lasing wavelength can be changed very rapidly by **adjusting/changing** the currents to the three electrodes.
- ○ The lasing **wavelength is controlled** by means of the currents to the three electrodes.

  日本語では,電流を制御して波長を調整するという書き方もあるが,英語での考え方は逆である:電流の調整によって波長を control する。すなわち,通常,「control」は,究極の目的または望みの結果に関して用いられる。

- ✗ *The roughness of the silicon surface needs to be <u>controlled</u> before oxidation.*

  粗さをcontrolすると,必ず粗さが小さくなるとは限らない。
- ○ The roughness of the silicon surface needs to be **reduced** before oxidation.

> **NOTE:** 変数をある値または二つの値の間に control するという言い方は適切ではない。この場合,その変数を set するか,change するか vary するように説明しよう。

- ✗ *The frequency can be <u>controlled from 9 to 11 GHz</u>.*
- ○ The frequency can be **varied between** 9 and 11 GHz.
- ○ The frequency can be **set to any value from** 9 to 11 GHz.

- ✗ *When the input power is large, the bias current is <u>controlled at a low level</u>.*
- ○ When the input power is large, the bias current is **set to** a low level.
- ○ When the input power is large, the bias current is **kept at** a low level.

tolerance

POINT:「Tolerance」の後ろの前置詞は通常「to」である。

- ○ This device exhibits good **tolerance to** thermal noise.
- ○ One problem is the lack of **tolerance to** a long series of consecutive identical digits.

respectively

POINT: 「**Respectively**」という単語は項目の出る**順番**に関して，日本語でいう「それぞれ」とまったく同じではない。

「Respectively」を使うには，**二つのリスト**が必要であり，最初のリスト内の項目の**順番**は二番目のリスト内の項目の**順番に正確に対応**している必要がある。したがって，リスト項目の順番を入れ替えても意味が変わらない場合は，「respectively」を使ってはいけない。

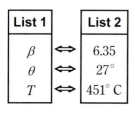

○ β, θ, and T have values of 6.35, 27°, and 451°C, respectively.
このリストの順番はとても大切である。以下の文で各リストには三つの項目があるが，変数リストと数値リストの順番は違うので，文は間違っている。

✗ β, θ, and T have values of 451°C, 6.35, and 27°, respectively.

Typical Mistakes

> 複数名詞はリストではない。

✗ The <u>two amplifiers</u> have input powers of <u>–10 dBm and 6 dBm</u>, respectively.
この文は，日本語では「それぞれ」を使ってよいが，英語では「respectively」を使ってはいけない。その理由は二つある。
  (a) 複数名詞はリストではない。
  (b) 「-10 dBm and 6 dBm」のリスト内の項目の順番は気にしていない。すなわち，「6 dBm and –10 dBm」で逆の順番で記述しても，文の意味は変わらない。
○ The two amplifiers have input powers of <u>–10 dBm and 6 dBm</u>.
○ The two amplifiers have input powers of <u>6 dBm and –10 dBm</u>.

> 一つの項目はリストにならない。

✗ The pulse widths were <u>13 ps</u> for the <u>pump</u> and <u>5 ps</u> for the <u>signal</u>, respectively.
○ The pulse widths were 13 ps for the pump and 5 ps for the signal.

NOTE: 「Respectively」が使える場合でも，以下の例のように，それを使わずに作文した方がもっと簡単でわかりやすく，さらに短くなる場合もある。

OK: The threshold current and output power are 1.5 mA and 70 mW, <u>respectively</u>.
BETTER: The threshold current is 1.5 mA, and the output power is 70 mW.

> **PRACTICE:** 次の文の中から「respectively」が正しく使われている文を選択せよ。

1. These graphs show Z versus X and Y, respectively.

2. Figure 6 compares the sensitivity curves of three resists. The pKa values in water are -20, -6, and -2, respectively.

3. For gaps of 30, 20, 15, and 10 μm, the finest resolution is 90, 80, 70, and 60 nm, respectively.

4. The threshold current was 3.2 mA for a 15-μm-diameter mesa and 6.5 mA for a 25-μm-diameter mesa. The maximum output powers were 0.85 mW and 2.0 mW, respectively.

φ	I_{th}	Max. P_{out}
15 μm	3.2 mA	0.85 mW
25 μm	6.5 mA	2.0 mW

5. The conditions were an applied pressure ranging from 139 to 417 Pa, and a vacuum of 10 Torr, respectively.

6. The red and blue lines are for devices 5 μm and 10 μm in diameter, respectively.

7. The exposure time is different for the first and second exposures, respectively.

8. Only an area with horizontal and vertical dimensions of 2.4 mm x 0.24 mm, respectively, can be illuminated.

9. The RF power and the cleaning temperature were kept at the optimum values of 220 W and 260°C, respectively.

10. The testing pads are placed on wires in the 1st, 2nd, and 3rd metal layers, respectively.

11. The static decision IC and the frequency divider IC cover operating ranges of 20 Gb/s and 20 GHz, respectively.

common vs. popular

DEFINITION: あることが **common** であるとは，それはよく起こる，よくある，普通に見かける，ありふれているという意味である．

- Green tea is a **common** drink in Japan.
- Passwords are the most **common** method of user authentication (ユーザー認証).
- The rate of oxygen consumption is a **common** measure of exercise intensity.
- Etching with a solution of H_2SO_4 and H_2O_2 is a **common** cleaning technique.

DEFINITION: あることが **popular** であるとは，それは多くの人々を楽しませて，人気があるという意味である．

- Green tea is a **popular** drink in Japan.
- Soccer is the most **popular** sport in Europe.
- Chocolate is always **popular** with children.

技術英語で，「**popular**」はほとんど使わない．

PRACTICE: 適切だと思われる「common」か「popular」，または両方を選べ．

1. Cellular phones are becoming very **common/popular** throughout the world.
2. Aluminum is a very **common/popular** material in silicon LSI fabrication.
3. Louis Vuitton is a very **common/popular** brand among young women.
4. Protocol LSIs and signal processing LSIs are very **common/popular** types of LSIs.
5. Beat Takeshi is a very **common/popular** TV personality (=テレビタレント).
6. Dogs are very **common/popular** pets in Japan.
7. This is one of the most **common/popular** methods of cleaning Si for LSI fabrication.

recently

POINT: 日本語の「最近」という言葉は，近い過去とこのごろの両方が含まれているのに対して，英語の「recently」は近い過去しか含まない。

DEFINITION: Recentlyに発生したことは，少し前の過去に発生したことである。

「Recently」に現在時制は使えない。

「Recently」は過去を指している。

- ✗ *Recently, many such algorithms <u>are reported</u>.*
- ○ Recently, many such algorithms **have been reported**.

新しいトレンドまたは今起こっていることに「recently」は使えない。

- ✗ *<u>Recently</u>, portable battery-operated devices <u>are</u> in widespread use.*
- ○ Portable battery-operated devices <u>are</u> **now** in widespread use.

- ✗ *<u>Recently</u>, lithium ion batteries <u>have been used</u> in hybrid cars in combination with a conventional engine.*
- ○ Lithium ion batteries **are now being used** in hybrid cars in combination with a conventional engine.

PRACTICE: 次の文に「recently」は正しく使われているか否か確認せよ。間違いがあればそれを直せ。

1. A big milestone (マイルストーン) was recently reached when...
2. Recently, these systems are investigated...
3. Recently, Dr. Sato asserted (断言する) that the cleaning solution used on production lines was not very clean.
4. Recently, the operating speed of CMOS LSIs reaches about 4 GHz.
5. We have recently moved to new research laboratories.
6. Recently, a great deal of attention is paid to these defects.
7. Recently, e-beam lithography was used to fabricate this circuit in 2014.

8. Recently, the significance of this phenomenon has been pointed out.
9. Recently, portable equipment uses low-voltage LSIs.
10. Recently, this method has been widely used.
    This method is the conventional one. (conventional = 現在一般的に使われている。Section 4の「**conventional**」の題を参照)
11. Until recently, we were unable to clarify the origin of these defects.
12. Recently, several organizations have developed their own steppers.
13. Recently, it is becoming more important to reduce costs.

simplified

POINT: 何かが **simplified**（簡略化した）と記述する場合，その元になっている，より **complicated**（複雑）なものがなければならない。

complicated

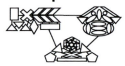

DEFINITION: 何かを **simplify** することは，その中の一部または細部を削除して，複雑さを減らすことである。

simplified

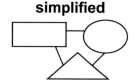

Good Examples

- Figure 3 shows a **simplified** version of the circuit in Fig. 2.
- The three-link robot can be treated as a two-link robot in the second stage. This **simplified** framework is employed to obtain a suitable combination of posture and energy.

simplified ≠ simple

これは **simple** な時計の図である。これは複雑なものから簡素化したものではないので，この場合，「simplified」を使ってはいけない。

Typical Mistakes

- To share photographs and other content, users desire a (~~simplified~~) **simple** way of transferring them quickly from a cell phone to other devices.
- Figure 6 is a (~~simplified~~) **simple** diagram of the system evaluation board.
- An acrobot is a (~~highly simplified~~) **very simple** model of a gymnast on a high bar.

introduce

> **POINT:** 技術英語では，「**introduce**」はよく間違って使われている。

Ex. 1: ✗ *To cook the food, we introduced a pan.*

> 日常英語で，この文は変に聞こえる。技術英語でも，この「introduce」の使い方は不自然である。

　○ To cook the food, we **used** a pan.

- Multichannel optical coupling (*was introduced*) **is used/employed** to connect the planar lightwave circuit to the InP chip.
- The ridge count（隆線数）(*was introduced*) **is used/employed** as an additional index for evaluating the quality of fingerprint images.
- We (*introduce*) **use/employ** a push-pull drive configuration to reduce the driving voltage.

Ex. 2: ✗ *We introduced cream to make the coffee taste better.*

> 前の例文と同じく，この例文も不自然である。

　○ We **added** cream to make the coffee taste better.

- A small hole was (*introduced*) **added/made** in the center of the defect cavity to change the resonant wavelength.
- Battery protection circuits were (*introduced*) **added** to prevent overcharging.
- Operating-point control electrodes were (*newly introduced*) **added** to the modulator.

Ex. 3: ➡ **explain, describe, discuss, review**

> 何かを説明するまたは描写するときに，「introduce」を使ってはいけない。

- Section 3 (*introduces*) **explains/describes** the device structure.
- The photoluminescence properties of graphene oxide are (*introduced*) **discussed/described** below.
- 口頭発表: Next, I'll (*introduce to you*) **explain** the theoretical model of our single-electron device.
- 口頭発表: Now, I'll (*introduce*) **review/explain** our previous work.

正しい使用法

「Introduce」の使い方はおもに二通りある。基本的なアイデアとして，いままでなかったものをある環境に新たに導入することである。

ある技術をはじめて使う場合

- To build over-100-GHz millimeter-wave systems with practical components, it was necessary to **introduce** photonic technologies **into** electronic systems.
- When optical lithography, which employs lenses, could not be extended any further, it was necessary to **introduce** X-ray lithography, which employs mirrors.

何かを完全に違うタイプのものに，新たに挿入する場合

Ex. 1: To introduce one substance or chemical component into a place or into a different substance
- It takes 1.5 min. to **introduce** the fluoro-compound **into** the chamber and raise the pressure to 4 MPa.
- One way to reduce the absorption at 13 nm is to **introduce** aromatic rings（芳香族環）**into** a polymer.

Ex. 2: To introduce a source of error
- In lithography, the use of off-axis incident light for exposure **introduces** another source of pattern placement error during printing.

Ex. 3: To introduce strain into a material
- This structure makes it unnecessary to **introduce** strong strain **into** the active layer to obtain polarization independence.

enough

POINT: 「Enough」を形容詞または副詞と一緒に使う場合，「enough」をその形容詞や副詞の後ろに置くこと。

✗ *The bandwidth is enough large.*
○ The bandwidth is **large enough**.

✗ *The tracking error does not converge enough fast.*
○ The tracking error does not converge **fast enough**.

Adjective Formation (-ing)

名詞 ＋ 現在分詞

The diode **emits light**.
①

能動態

②
It is a **light-emitting** diode.

③　light ← emits ← diode

① 「Diode」は文の主語である。

② 現在分詞（–ing）は能動態を示す。

③ この形式の文を理解するには，逆方向に読めばよい。

EXAMPLES:

The design <u>saves energy</u>.　　⇨　　It is an **energy-saving** design.
The work <u>consumes time</u>.　　⇨　　It is **time-consuming** work.

PRACTICE: 次の文を完成せよ。

PART A

1. The mechanism holds a wafer. What kind of mechanism is it?
   It is a _____.
2. The copper contains zirconium. What kind of copper is it?
   It is _____.
3. The performance leads the industry. What kind of performance is it?
   It is _____.

PART B

Example: What is a light-emitting diode?
   It is *a diode that emits light.*

1. What is dispersion-reducing fiber?
   It is _____.
2. What are noise-reducing techniques?
   They are _____.
3. What is wire-bonding equipment?
   It is _____.

Adjective Formation (-ed)

名詞　+　過去分詞

The system is **powered by a battery**. ①
①
受動態
②
a **battery-powered** system
③
battery ← **by** ← powered ← system

① 「System」は文の主語である。

② 過去分詞(-ed)は受動態を示す。

③ この形式の文を理解するには、**逆方向**に読みながら、ハイフンの代わりに適切な前置詞を入れる。

EXAMPLES:
The mountain is <u>covered with snow</u>.　⇨　It is a **snow-covered** mountain.
The engine is <u>cooled by water</u>.　⇨　It is a **water-cooled** engine.

PRACTICE　A: 次の文を完成せよ。

PART A
1. The design is aided by a computer. What kind of design is it?
   It is _____.
2. The silicon is doped with boron. What kind of silicon is it?
   It is _____.
3. The pulse is generated by a laser. What kind of pulse is it?
   It is _____.

PART B
Example: What is a battery-powered system?
It is *a system that is powered by a battery.*
1. What are hand-made shoes?
   They are _____.
2. What are semiconductor-based components?
   They are _____.
3. What is a speed-oriented circuit technique?
   It is _____.

Adjective Formation (-ed)

> **PRACTICE B:** 次の問に答えよ。下線の引かれた単語が主語でない場合, まず文を受動態に変えてから問に答えよ。
>
> **Ex. 1:** This <u>appliance</u> saves time. What kind of appliance is it?
> It is a **time-saving** <u>appliance</u>.
>
> **Ex. 2:** Oil fuels the <u>power plant</u>. What kind of power plant is it?
> まず, power plant を主語にする。
> The <u>power plant</u> is fueled by oil.
> It is an **oil-fueled** <u>power plant</u>.

1. The <u>resist</u> is rinsed with water. What kind of resist is it?

   _____

2. The <u>network</u> switches packets. What kind of network is it?

   _____

3. Liquid fills the <u>section</u>. What kind of section is it?

   _____

4. The <u>power</u> is averaged over time. What kind of power is it?

   _____

5. This <u>structure</u> blocks the current. What kind of structure is it?

   _____

6. This <u>equipment</u> sorts letters (in a post office). What kind of equipment is it?

   _____

7. The <u>device</u> was grown at a low temperature. What kind of device is it?

   _____

8. The user specifies the <u>parameters</u>. What kind of parameters are they?

   _____

9. *Chemistry:* This <u>factor</u> limits the rate (of the reaction). What kind of factor is it?

   _____

compensate: 他動詞 vs. 自動詞

他動詞: compensate X　（金で償う）

DEFINITION: AさんがBさんに傷害，紛失，破損，不便などを負わせた場合，AさんがBさんに **compensate** するとは，Bさんの損害のための補償で，AさんがBさんに金を払うことである。

- We <u>compensated him</u> for his trouble.
- The company <u>compensates employees</u> who work in foreign branches.

技術英語では，他動詞の方はほとんど使われていない。

自動詞: compensate for X　（埋め合わせる）

DEFINITION:「X **compensates for** Y」というのは，XがYの悪影響を打ち消すことを意味する。

- ✗ Forward error correction is an effective way to <u>compensate the degradation</u> in signal quality.
- ○ Forward error correction is an effective way to **compensate for** the degradation in signal quality.
- If the positional error of the MEMS mirror becomes large, stabilization control cannot **compensate for** it.
- The amplifier has auto offset control to **compensate for** the voltage difference between the two inputs.
- An amplifier **compensates for** the loss.

conventional

DEFINITION: Conventional な方法，製品，慣習などは，現実の世界で長い間使用されてきて，今でも使用されて，通常のタイプと見なされるものである。

- LED lamps（LED照明）last longer than **conventional light bulbs**.
  conventional light bulb = 白熱電球
- Hydroponics（水耕栽培）produces higher yields than **conventional farming**.
  Conventional な農業では，植物を土壌で繁殖させる。
- Quantum computers will be able to solve certain types of problems much faster than **conventional computers** can.
  Conventional なコンピュータは，デジタルデータを処理するためにトランジスタを使用する。

研究所にしか存在していないものには「conventional」を使ってはいけない。

Punctuation: Comma 1   87

Comma 1

1. 主節から単語, 句, および節の分離

Beginning: XXXXX, ――主―節――.

- First, a 6" Si wafer is thermally oxidized.
- For an outdoor experiment, portable equipment is needed.
- To handle video, the data rate must be at least 10 Gb/s.
- If the sample is electrically conductive, a current can pass through it.

Middle: ――主――, XXXXX, ――節――.

- There are, however, problems to be solved.
- Visible light, on the other hand, has a shorter wavelength than radio waves.
- A lower capacity, indicating that storage changed the properties, was observed.
- The transition time, which is the charging time, is very long.

End: ――主―節――, XXXXX.

- Only two devices are used as circuit elements, greatly simplifying the fabrication process.
- The refractive-index contrast is very high, with Δ being about 40%.
- The particle size has a random distribution, centered on a diameter of 5 μm.

2. リストの項目分け

- The key issues are the detection, repair, and reduction of defects.

NOTE:　通常, リストの項目は二つ以上あれば, 「and」の前にコンマを入れることは任意であり, 文章全体でスタイルを統一すればよい。しかし, 次の例のように, コンマが不可欠である場合もある。

- The three layers consist of InGaAs, InGaAsP and InGaAs, and InP.

コンマがなければ, 二番目と三番目のレイヤは何によってできているか明らかではない。

| InGaAs |
| InGaAsP & InGaAs |
| InP |

3. 等位接続詞で接続している二つの主節の分割

(等位接続詞: and, but, for, nor, or, yet, so)

- One antenna is an emitter, **and** the other is a detector.
- It has a large capacity, **but** it also consumes a great deal of power.
- This method does not require the construction of an inverse model, **yet** it provides satisfactory compensation.

DANGER:　二つの主節の間に等位接続詞がない場合, コンマだけで接続してはいけない。この場合, 主節を二つの文に分解するか, またはセミコロンで接続するようにする。

✗ F is a low-pass filter, it eliminates high-frequency noise.
○ F is a low-pass filter. It eliminates high-frequency noise.
○ F is a low-pass filter; it eliminates high-frequency noise.

> **PRACTICE:** 次の文にコンマを入れよ。読みやすくするために，一部の主節に下線が引いてある。

1. Unlike radio waves terahertz waves are not scattered very much by dust soot or smoke.
2. The sensor unit weighs only 5 g so it is light enough to be unnoticeable by a wearer.
3. Table 2 compares length area and execution time.
4. Direct bonding is an easy way to put a laser on Si but there are two drawbacks.
5. Consider as an example an experiment in which a coin is tossed three times.
6. The use of passive couplers provides a small phase error but it makes the loss large.
7. All cases used 0.75-μm-thick PMMA resist with the results being an average for isolated lines isolated spaces and line-space arrays.
8. Then it gradually becomes paler and finally turns grayish.
9. The n-type dopant is silicon and the p-type is carbon.
10. On the other hand if X is low <u>Y will be low</u> after the clock goes high.
11. However if these constraints are present <u>problems arise</u> if one attempts to simplify the timing graph.
12. The overhead is 4 cycles so the total number of cycles is 68.
13. If we use a wavelength filter as a multiplexer the loss is low but the size is large.
14. With image placement targets as low as 35 nm <u>all contributions must be minimized</u> including that from the e-beam system and process-induced distortion.
15. Common layout procedures such as symbolic layout and layout compaction destroy the symmetry of critical analog layouts impacting performance.
16. When the clock input is low <u>transistors P_3 and P_4 act as resistive loads for the first stage</u> which acts as a linear amplifier for small input swings and as a swing limiter for large input swings.
17. <u>High voltage reduces scattering</u> resulting in better resolution straighter side walls and smaller proximity effects.

Unnecessary Words 1

Ex. 1: A dog is an animal.
　次のことはいえるか？
- I have a <u>dog animal</u>.

答えは「ノー」である。だれでも犬が何かを知っているため、「animal」という言葉は不要である。

Ex. 2: Reactive-ion etching is a technique.
　さて、次の文を吟味しよう。
- The waveguide is formed by the <u>reactive-ion-etching technique</u>.

「Technique」という言葉は本当に必要であるか？まず、それを削除すると文の意味はどのように変わるか見てみよう。
- The waveguide is formed by reactive-ion etching.

意味はまったく変わらない。すなわち、ここで「technique」という単語は**不要**である。このような間違いは日本語を英語に直訳する場合によく生じる。以下の単語は、表現の後ろによくつけられるが、英訳する場合、不要な確率が高いものである。

effect, operation, process, technique

これらの単語を用いる必要があるかどうかを判断するには、まずそれを削除して意味が変わるかどうかテストしてみることである。

特に「effect」の使い方に注意

次の文を吟味しよう。
- ✗ *After the Si is cut, <u>the oxidation effect</u> produces a thin layer of SiO_2 on the surface.*

酸化と酸化の影響の間には大きな違いがある。**酸化**は化学反応である。
- The <u>oxidation</u> of Si produces SiO_2.

酸化の一つの影響は、材料の特性を変えることである。
- One of <u>the effects of the oxidation</u> of a Si surface is an increase in the electrical resistance.

- ○ After the Si is cut, **oxidation** produces a thin layer of SiO_2 on the surface.

PRACTICE: 下線の引かれた単語は正しく使われているかどうか検討せよ。

1. The shielding <u>effect</u> provided by the 3D structure enables a variety of useful devices to be fabricated.
2. To monolithically integrate the devices, we employ <u>a</u> regrowth <u>technique</u>.
3. ...<u>the</u> quantum-confined Stark <u>effect</u> comes into play.

Section 4

4. Anisotropic Si etching <u>technology</u> is one of the most important technologies for bulk micromachining <u>processes</u>.
5. The collimation <u>effect</u> is caused by the flaring of the potential boundary.
6. One critical problem is short-channel <u>effects</u>, such as a threshold voltage shift.
7. This structure is essential for achieving low-threshold-current, high-output-power <u>operation</u>.
8. <u>An</u> offset-canceling <u>technique</u> is employed in the limiting amplifier.
9. Two special techniques were used for the fabrication: <u>an</u> image reversal <u>process</u> and <u>an</u> oxidation <u>process</u>.
10. High speed results from the suppression of <u>the</u> current blocking <u>effect</u>.
11. The interconnection lines are formed by lift-off and <u>an</u> air-bridge <u>technique</u>.
12. The p+ regions are formed by Zn diffusion during <u>the</u> alloying <u>process</u>.
13. This device employs a Coulomb blockade <u>effect</u> to manipulate individual electrons.
14. This type of dependence suggests that the Auger <u>effect</u> is involved.

Prepositions 4

空欄を適切な前置詞で埋めよ。必要がなければ、×で埋めよ。

REVIEW

a. X is responsible _____ Y.
b. X is identical _____ Y.
c. This paper concerns ____ X.
d. X is suitable _____ Y.
e. _____ the other hand, ...
f. X equals _____ Y.
g. X is independent _____ Y.
h. X is capable _____ Y.
i. X is ____ the order ____ 10^6.
j. X influences _____ Y.

CHECK YOUR KNOWLEDGE

1. This paper is concerned ____ X.
2. X is added _____ Y.
3. X is related _____ Y.
4. X takes Y _____ account.
5. X is connected _____ Y.
6. an increase _____ the length
7. X is comparable _____ Y.
8. There is no information ____ Y.
9. X is sandwiched _____ Y and Z.
10. Sato et al. mention _____ two problems ...

Section 5

effective

has been used vs. is used

number

by vs. with

the both, the each, the another...

Keep Related Words Together

multi-

fixed

coincide

traditional

correspond

Punctuation: Comma 2

Style: Unnecessary Words 2

Prepositions 5

effective

> **DEFINITION:** ある手法が effective であるとは，その手法が望みの結果を生み出すことを指す。

　ある手法が effective であるとは，その手法はベストという意味ではなく，その手法を使って望みの結果が得られることだけを意味する。例えば，火をたきつける場合，マッチ，ライター，拡大鏡（天気のよい日）または木片に棒をこすることによって生じる摩擦などが使える。これらの方法は，よし悪しがあるものの，すべて effective である。

effective の使用法

be effective

- This control method **is** simple and **effective**.

be effective in ___ing

- The polishing **is effective <u>in making</u>** the surface flat.
- The system **is effective <u>in reducing</u>** production costs and CO_2 emissions.
- The results show **how effective** the isolation walls **are <u>in reducing</u>** the interference.

be an effective way/method to do…
be an effective way/method of doing…

※この「to」と「of」は「effective」と全然関係せず，「way」か「method」と関連している。

- Our new process is **an effective <u>way to remove</u>** the residue.
- **One of the most effective <u>ways to lower</u>** the power consumption is to reduce the supply voltage.
- An adiabatic charging circuit is **an effective <u>way of storing</u> energy** in a supercapacitor.
- **An effective <u>method of improving</u>** the characteristics is to shorten the gate length.
- Sacrificial oxidation and annealing in hydrogen **are effective <u>methods of reducing</u>** the defect density.

POINT: 「Effective」を使った構文には多くの間違いが見られる。次の構文は正しい英語の表現ではないので、気をつけよう。

　　　× ~~be effective to do~~　　　× ~~be effective for doing~~

× *This technique is effective to improve the performance.*

× *This technique is effective for improving the performance.*

○ This technique **is effective in improving** the performance.

○ This technique **is an effective way to improve** the performance.

effective vs. efficient

DEFINITION: ある手法が **efficient** であるとは、その手法が時間、エネルギー、原料などを節約することを意味し、すなわち、効率的であることを指す。

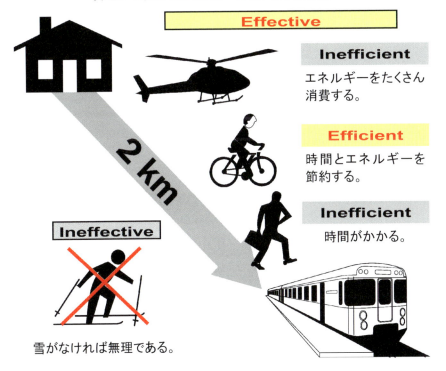

自宅から駅まで行くのに一番よい方法は何か？

has been used vs. is used

POINT:「**Has been used**」とは現在ではなく，過去を指している。

OK ① In the QPSK transmission experiments reported **so far**, two types of modulators **have been used**.

✗ *In this experiment, a 1.55-μm DFB laser has been used to generate the input signal.*
この実験は，過去の特定の時刻に行ったため，過去形で説明する必要がある。

② In this experiment, a 1.55-μm DFB laser **was used** to generate the input signal.

✗ *Lithium ion batteries have been used in our daily lives to power portable electronic devices.*
この文は，現在の状況を説明しているので，現在時制を使用する必要がある。

③ Lithium ion batteries **are used** in our daily lives to power portable electronic devices.

Typical Mistakes

✗ *In general, a voltage of 3.3 V has been used for I/O circuits.*
この文を読むと，3.3 Vの電圧はいままで使用し，これからは違う電圧に変わると誤解される。

○ In general, a voltage of 3.3 V **is used** for I/O circuits.
通常3.3 Vの電圧が使われている。

✗ *In conventional CDR circuits, an external variable-delay line has been used to adjust the clock timing.*
「**Conventional**」という言葉の指すものは，一般的に使われているものである。すなわち，現在のものである。[Section 4の「**conventional**」の題を参照]

○ In conventional CDR circuits, an external variable-delay line **is used** to adjust the clock timing.

✗ *The knife-edge method has been used to measure the diameter of an electron beam.*
○ The knife-edge method **is commonly used** to measure the diameter of an electron beam.

has been used vs. is used

Good Examples

- **So far**, this system **has been used** to test more than 10 actual chips.
- Photonics and optical-fiber technologies **have so far been used** mainly for telecommunications systems.
この文は，将来この技術が通信システム以外にも使われるだろうということをほのめかしている。
- UHV scanning transmission microscopy (STM) **has been used for** the in-situ analysis of atomically flat GaAs surfaces. **However**, ex-situ atomic force microscopy (AFM) **is now** the method of choice for this kind of examination.
この文は，**いままで**UHV STMが使われてきたが，**現在**はAFMが好まれる方法であることを意味する。

PRACTICE: 正しい文を選択せよ。また，間違いの文を訂正せよ。

1. A lot of multimode fiber <u>has been installed</u> for LANs in office buildings.

2. A lot of multimode fiber <u>has been used</u> for LANs in office buildings.

3. Some of these ICs <u>have already been used</u> in 40-Gb/s transmission experiments.

4. Three main methods <u>have been used</u> to grow crystals of organic materials.

5. Wavelength-division multiplexing systems <u>are now being used</u> in metropolitan and access networks.

6. The liquid crystal method <u>has been used</u> for hot-spot analysis for a long time, but now something better is needed.

7. GaAs FETs <u>have been widely used</u> in circuits operating at microwave frequencies.

8. Organic solvents <u>are widely used</u> for the electrolyte of Li-ion batteries.

9. The amplifier we developed <u>has been used</u> in the transmitter and receiver of a 120-GHz-band wireless system.

10. Two types of power supply systems <u>are used</u> in telecommunications facilities.

11. Conventionally, lead-acid batteries <u>have been used</u> for backup power supplies.

number

POINT:「**X number**（X番号）」と「**number of Xs**（Xの数）」という表現を混同してはいけない。

X number（X番号）

　数字はよく区別のために使用される。例えば、ホテルの各部屋は別々の番号によって区別される。それは **room number**（部屋番号）と呼ばれ、部屋のドアと鍵に書いてある。以下は、識別番号の例を示す。
license plate number（自動車登録番号）
registration number（登録番号）
serial number（シリアル番号）
sample number（サンプル番号, 試料番号）

number of Xs（Xの数）

　何かの数がいくつあるかを指すために使用される。例えば、右に示すホテルの図において、**the number of rooms**（部屋の数）は9である。それに対して、赤いボックスで示す部屋の **room number** は203である。

room number ≠ number of rooms

例：
- As the **channel number** increases, the wavelength becomes smaller.

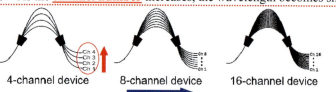

- As the **number of channels** increases, fabrication becomes more difficult.

Typical Mistakes

✗ *The electrons number is about 100.*
○ **The number of electrons** is about 100.

✗ *For this chip, the total gate number is 780.*
○ For this chip, the total **number of gates** is 780.
○ For this chip, the total **gate count** is 780.

右の図について，次の文があったとする。
- The problem is how to reduce the terminals.
どういう意味だろうか？これを書いた人は以下の(a)か(b)か，どちらのつもりであるかわからない。

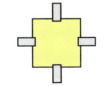
Terminal（端子）

(a) より小さい端子　　(b) より数の少ない端子

このような意味不明な文は技術英語の悪い例である。
(a)の場合，正しい英語は以下である。
- The problem is how to reduce the size of the terminals.

(b)の場合，正しい英語は次である。
- The problem is how to **reduce the number** of terminals.

Reduce the terminalsができない理由は，端子の概念自身には数量が伴わないからである。しかし，端子の重さ，コスト，密度，厚さ，電気抵抗などは数量が伴うため，それを減らすことができる。また，もちろん，**端子の数も減らす**ことができる。

> 物事自身を増減することはできない。増減できるのはそのものの何らかの数量に伴う部分に限る。

Typical Mistakes

✗ Using the combined instruction *reduced the steps* in the procedure from 58 to 33.
○ Using the combined instruction **reduced the number of steps** in the procedure from 58 to 33.

✗ *The increase in the photoluminescence intensity is due to the decrease in the defects in the quantum wells.*
○ The increase in the photoluminescence intensity is due to **the decrease in the number of defects** in the quantum wells.

✗ *The loss gets larger as the channels increase.*
○ The loss gets larger as **the number of channels increases**.

✗ *The bit error rate is the error bits divided by the total bits.*
○ The bit error rate is **the number of error bits** divided by **the total number of bits**.

Section 5

by vs. with

POINT:「で」の訳し方に注意。

この手紙はペンで書かれた。　　This letter was written **with** a pen.

with + 具体的なもの

道具, 器具, 物体または物質を使って何かをする場合,「**with**」を使う。

- The temperature was measured **with** a thermometer.
- The substrates are rinsed **with** water.
- The surface was observed **with** a scanning electron microscope. 　器具

by + 抽象的なもの

方法, 手順, プロセス, テクニックなどを使って何かをする場合,「**by**」をよく使う。

- The buffer layer was grown **by** molecular-beam epitaxy.
- The coordinates are calculated **by** simple addition.
- The gate patterns are extracted **by** procedure P1.
- The surface was observed **by** scanning electron microscopy. 　技術

例外:

- The devices are fabricated **with** silicon technology.

「Make with」と「fabricate with」は製造プロセスを表すのに使う場合もある。この使用法は上の議論と関係ない。

次のペアを比較しその違いを吟味して, 文法の面から理解を深めよう。
- Describing what you do:
  - 能動態　We generate the pulses with a fiber laser.
  - 受動態　The pulses are generated with a fiber laser (by us).
- Describing what happens:
  - 能動態　A fiber laser generates the pulses.
  - 受動態　The pulses are generated by a fiber laser.

この文では,「by」は受動態の動詞に関係があり, 上の議論の限りではない。

PRACTICE: 空欄を「by」か「with」で埋めよ。

1. The surface morphology was observed _____ a microscope.
2. The pulses are encoded _____ a modulator.
3. They were fabricated _____ the following procedure.
4. This is an image produced _____ projection.
5. It is difficult to control the wavelength _____ a conventional circuit.
6. During a measurement, we illuminate the region _____ a laser beam.
7. The copper is removed _____ an HF solution. （フッ化水素の水溶液）
8. This chip was designed _____ the standard-cell method.
9. Thin layers of WSi and Au are formed _____ sputter deposition, and a Au side wall is made _____ electroplating.
10. The wafer is cleaved _____ an automatic cleaving machine.
11. We measured the performance _____ on-wafer probes.

the both, the each, the another...

POINT: 以下の表現は正しい英語ではない。

✗ the both
✗ the some
✗ the each

NOT ENGLISH!

✗ the all
✗ the another

- These results demonstrate that (~~the~~) **both** transmitters emitted high-quality optical signals.
- The loss for (~~the~~) **both** polarizations is about 9.5 dB.
- Irradiation with infrared light causes (~~the~~) **some** parts of the molecule to vibrate.
- A different modulator is used for (~~the~~) **each** optical path.
- It is necessary to connect (~~the~~) **each** subcell by means of a tunneling junction.
- (~~The all~~) **All the** Telcordia-0468 tests were conducted on 11 modules.
- (~~The all~~) **All the** bitlines are set to a voltage of $3V_{DD}/4$.
- (~~The~~) **Another** problem is how to connect Si and silica waveguides.

Keep Related Words Together

Broken Adjective Clause（形容詞節）

 Section 1の「**Adjective Clause: Short Form**」の題で述べたように，形容詞節の短縮形を分解してはいけない。すなわち，その一部を名詞の前において，残りの部分を名詞の後ろにおいてはいけない。

✗ The <u>obtained</u> results <u>by our new method</u> agree well with the calculation results.

✗ <u>Deposited</u> film <u>by sputtering</u> was examined.

正しい文は以下のとおりである。

◯ The results <u>obtained by</u> our new method agree well with the calculation results.

◯ Film <u>deposited by</u> sputtering was examined.

Method　　Problem

次の正しい文を吟味しよう。

◯ …the fabrication of lasers…

「Fabrication」の後ろに「*method*」という言葉を入れると以下のようになる。

✗ …the *fabrication method of lasers*…

この場合，この表現の基本的構造は「**method of lasers**」となることに気をつけよう。レーザーは手法を持たないため，上の文は意味をなさない。「Method」を挿入することにより「fabrication」と「of lasers」との間の関連は途切れている。正しい表現は以下の通りである。

◯ …a method of fabricating DFB lasers…

◯ …a fabrication method for DFB lasers…

「**Problem**」という言葉の場合も，同様の間違いはよく見られる。

✗ We consider the *rejection problem for a disturbance*, $w(t)$, that satisfies $\Delta w(t) \neq 0$.

◯ We consider the problem of rejecting a disturbance, $w(t)$, that satisfies $\Delta w(t) \neq 0$.

Too Far Apart

次の正しい文を吟味しよう。

✗ A power-flow control system was designed that provides good tracking performance and compensates for disturbances <u>based on the EID method</u>.

何がEID methodに基づいているか？それはdisturbance compensation（外乱補償）であるか，またはtracking（追跡）とdisturbance compensationであるか？答えはnoである。実際には，設計がEID methodに基づいている。しかし，「design」と「method」という単語は，あまりにも遠く離れているため，この文が非常に理解しにくく，よい構文ではない。

Keep Related Words Together   101

○ The EID **method** was used to **design** a power-flow control system that provides good tracking performance and compensates for disturbances.

Typical Mistakes

✗ A **protection** method **against** shocks during fabrication is needed.
○ A method of **protecting against** shocks during fabrication is needed.

✗ This effect becomes stronger as the number of **coupled** quantum dots **to the quantum wire** increases.
○ This effect becomes stronger as the number of quantum dots **coupled to the quantum wire** increases.

✗ This paper concerns a **capacity** estimation method **of** Li ion batteries.
○ This paper concerns a method of estimating the **capacity of** Li ion batteries.

PRACTICE: 関連の途切れた次の表現を直せ。

1. an estimation method of the current

2. injected electrons into the base

3. the waveform of transmitted signals over our new microstrip line

4. We have developed a new trimming technique of planar lightwave circuits.

5. Stored data is destroyed at a voltage of 1.8 V in a 1-V SRAM.

6. to develop a design method of DC power networks

7. an integration technique of transistors and photodiodes

8. a replicated pattern by ECR plasma etching

9. The applied bias voltage to the device was −1.0 V.

10. the estimation problem for the normalized longitudinal force

multi-

「**Multi-**」は接頭辞であり，独立した単語ではない。

× *This receiver has <u>multi</u> channels.*
「Multi」は単語ではないため，形容詞として使えない。この場合「**multiple**」を使うべきである。
○ This receiver has **multiple** channels.

ほとんどの「**multi-**」で始まる単語は形容詞である。

× *This receiver has <u>multichannels</u>.*
「Multichannel」は形容詞なので，主語または目的語として使えない。また，その複数形も存在しない。
○ This is a **multichannel receiver**.

例外―名詞: multimedia, multifoil, multilayer, multimillionaire, multinational, multinomial, multiplet, multiplication, multiplier, multiplicand, multiplicity, multitude, multiversity
例外―動詞: multiply, multitask

「**Multi-**」の後ろの文字が母音であり，かつハイフンの挿入により単語の発音がわかりやすくなる場合，ハイフンを挿入しよう。

NO HYPHEN: multilevel, multistage, multiaddress
HYPHEN: multi-electrode

「**Multi-**」は名詞または形容詞と組み合わせることができる。辞書でその正しい形をチェックしよう。

multi- + 名詞: multiaddress, multilane, multilevel, multistage
multi- + 形容詞: multicolored, multicultural, multidirectional, multilingual
両方: multifunction—multifunctional, multilayer—multilayered

PRACTICE: 「Multi-」で始まる単語を使って，次の文を完成せよ。

1. What kind of device has several electrodes?
   A _____ does.
2. What kind of module holds several chips?
   A _____does.
3. What kind of circuit can perform many functions?
   A _____ can.
4. What kind of structure has several layers?
   A _____ does.
5. What kind of system has several processors?
   A _____ does.

fixed

POINT: 一般的に「fixed」は何かが動かないことを示すために使用される。すなわち，そのものは調整や変更の対象ではない。動いたり，調整や変更がされたりするものとの相違を明らかにするために，「fixed」はよく使用される。

- A pair of glass wedges was placed in front of the objective lens. The position of one was **fixed**, and the other was **movable**.

- In an arrayed-waveguide grating, the wavelengths are **fixed** and cannot be **tuned**.

- The capacitance between the **deformable** upper electrode and the **fixed** lower electrode was calculated.

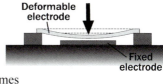

- **fixed** charges vs. **mobile** charges
- **fixed-length** frames vs. **variable-length** frames
- a **fixed-wavelength** filter vs. a **tunable-wavelength** filter

Typical Mistakes

変数の数値を単に記すためには「fixed」ではなく，be-動詞を用いること。

- The oxygen pressure (~~was fixed at~~) **was** 400 mTorr.
- The length of the payload (~~is fixed to~~) **is** 1,200 bytes.
- The intensity of the main pulse (~~was fixed to~~) **was** 1.3×10^{16} W/cm^2.

変数に数値を与えることを記すためには「fixed」ではなく，「set」を用いること。

- The band gap wavelength, λ_g, (~~was fixed to~~) **was set to** 1.39 µm.
- The integration time for the lock-in amplifier (~~was fixed to~~) **was set** to 1 ms.
- The total thickness of the layers (~~was fixed to~~) **was set to** half the wavelength.

変数の数値が変わらないことを記すためには「fixed」ではなく，「constant」を用いること。

- The current injected into the SOA **was constant**. = A **constant current** was injected into the SOA.

- β has **a constant value** of around 10.
- **A constant** D-FF **output** indicates that the PLL is in the locked state.
- The conditions were **a constant gate bias** of 0 V and a drain voltage ranging from 1.0 V to 1.8 V.

> 一つのものが別のものに結びつけられていることを記するために、「fix」ではなく、「attach」または「mount」という言葉を用いること。

- In this probe, the EO crystal (~~is fixed to~~) **is attached to** the tip of an optical fiber.
- All the parts (~~are fixed in~~) **are mounted in** the package.
- In a conventional package, an RF circuit board (~~is fixed to~~) **is attached to** the side of the chip.
- The housing (~~was fixed to~~) **was mounted on** a metal substrate.

> **PRACTICE:** 正しい文を選択せよ。また、間違いの文を訂正せよ。

1. In our model, the thickness of a metal line is fixed to 0.5 μm.
2. Tunable lasers provide a good backup for fixed-wavelength lasers.
3. In this experiment, the received power was fixed to -20 dBm.
4. The heat flow profile inside the device changes, even though the total power dissipation is fixed.
5. We used fixed-length optical packets to keep the control electronics simple.
6. The shielding protects the fibers fixed to the plug.
7. In the simulations, the input frequency was fixed at 2 MHz.
8. The number of "1" bits in each packet was fixed.
9. The frequency was fixed at 2 GHz.
10. Four fiber-lens assemblies are fixed on the metal block.
11. The timing conditions for the clock and data are fixed inside the chips.
12. The current injected into the active region was fixed at 120 mA.

coincide

POINT: 二つの物事が **coincide** するとは，それらは同じということではない。

DEFINITION:
1. あるイベントが他のイベントと **coincide** するとは，この二つのイベントが同時に起こることを指す。
2. 二つのものが **coincide** するとは，この二つのものは同時に同じ空間を占めていることを指す。

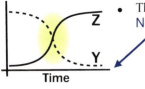

- The sharp drop in Y **coincides with** the rapid rise in Z. Note that the horizontal axis of the graph is time.

この単語は以上のような特別な意味を持つため，技術英語では使う場面は少ない。

coincide → agree with / be in good agreement with / be equal, be the same as / be consistent with

- These results (~~coincide with~~) **agree with** those results.
- These results (~~generally coincide with~~) **are in good agreement with** the results obtained by the back-to-back test.
- This temperature (~~coincides with~~) **is about the same as** that derived from the curve in Fig. 5.
- This result (~~coincides with~~) **is consistent with** Gaussian beam theory.

traditional

POINT: 「Traditional」という単語は，技術でなく，おもに文化に関連しているので，技術英語ではあまり使わない。

Good Examples
- ○ traditional Japanese architecture
- ○ traditional Chinese medicine
- ○ traditional German songs

Typical Mistakes ~~traditional~~ → **conventional**

- A smart grid differs from a (~~traditional~~) conventional power grid in several important ways.
- The integrated model provides higher accuracy and smaller variations than (~~traditional~~) conventional optimization methods.
- Network-on-chip architecture employs packet transmission instead of (~~traditional~~) conventional bus communications.

correspond

POINT 1: 「Correspond」は，異なるタイプの二つのものが類似または同等の機能，構造，位置，量などを有することを示すために使用される。

Good Examples

Foreleg（前肢）	⇔	Arm
Front paw	⇔	Hand
Tail	⇔	X

- A cat's <u>foreleg</u> and <u>front paw</u> **correspond to** a human <u>arm</u> and <u>hand</u>, respectively; but the tail does **not correspond to** anything.

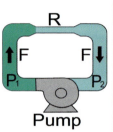

Pump	⇔	Battery
Pressure	⇔	Voltage
Vol. flow rate	⇔	Current
Poiseuille's Law $F = \dfrac{P_1 - P_2}{R}$	⇔	Ohm's Law $I = \dfrac{V_1 - V_2}{R}$

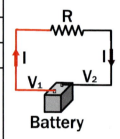

- The <u>pump</u> in a water circuit **corresponds to** the <u>battery</u> in an electrical circuit. In addition, pressure **corresponds to** voltage, volume flow rate **corresponds to** current, and Poiseuille's Law **corresponds to** Ohm's Law.

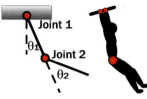

- The <u>first and second joints</u> of an acrobot **correspond to** a gymnast's <u>hands and hips</u>, respectively

- In a digital signal, the <u>high</u> level **corresponds to** <u>one</u>, and the <u>low</u> level **corresponds to** <u>zero</u>.

- In a strong wind, the largest <u>drop in received power</u> was 1.2 dB, which **corresponds to** about a 0.2° <u>rotation</u> of the antenna.

- A <u>resonance frequency</u> of 300 GHz **corresponds to** a <u>crystal thickness</u> of around 0.2 mm.

POINT 2: 二つの変数か数値などが同じまたは同等であり，または二つの結果が一致することを示すために，「correspond」はあまり使用しない。特に，be-動詞，「equal」，「be equivalent to」，および「agree with」を使うことができるならば，「correspond」は使わない方がよい。

- $3V_{DD}/4$ (~~corresponds to~~) **is** the precharge voltage.
- The difference between the intensities of the two signals (~~corresponds to~~) **is** the DC crosstalk.
- This voltage (~~corresponds to~~) **is** the voltage at which the photocurrent starts to increase rapidly.
- The calculation results (~~correspond to~~) **agree well with** the experimentally obtained results
- The measured terahertz absorption bands (~~correspond well to~~) **agree well with** the calculated frequencies.
- The typical amount of data (~700 MB) (~~corresponds to~~) **is equivalent to** about 140 photographs taken with a 10-megapixel camera.

POINT 3: 図の中のドット，カーブ，写真などを説明するために，「correspond」ではなくて，**be-動詞**，「**be for**」，「**indicate**」，「**represent**」，「**show**」などの言葉を使用すること。

- The dots in Fig. 7 (~~corresponds to~~) **are** measured values, and the curve was calculated.
- The images in Fig. 6 (~~correspond to~~) **show** the surface morphology of the InGaAs channel.
- The solid line in Fig. 9 (~~corresponds to~~) **indicates (shows, represents)** the measurement results for 26 g of charcoal.
- In Fig. 6, the yellow and red bars (~~correspond to~~) **are for** 10- and 1-Gb/s devices, respectively.

Comma 2

1. 記号の導入

… the threshold voltage**,** *Vth***,** …

この場合，括弧を使ってもよい：

- …the threshold voltage (*Vth*)。

ただし，文章全体で同じスタイルで統一すべきである。

コンマは記号の導入にいつも使うとは限らない。記号を省略しても文が成り立つときは，コンマを入れる必要がある。

Ex. 1:   Substrate effects degrade the quality factor, Q.   [OK]

ここでQを省略すると

- Substrate effects degrade the quality factor.   [OK]

この二番目の文は正しいから，一番目の文のシンボルの前に**コンマが必要で**ある。

Ex. 2:   Consider a device of length L.   [OK]

ここでLを省略すると

- Consider a device of length.   [No good]

この二番目の文は意味が通じないから，一番目の文のシンボルの前に**コンマを入れてはならない**。

2. 「For example」の前後

- This link is suitable for use as a temporary fixed wireless link**,** for example**,** for the transmission of material for TV broadcasting.

3. 数式と後ろのwhereとの間

- The input nonlinearity, Ψ, is decomposed into two parts:

$$\Psi(u(t)) = u(t) + d(u(t))\text{,} \qquad (4)$$

where $u(t)$ is the linear part and $d(u(t))$ is the rest.

4. インライン（行中）の住所の部分の分割

ABC Corp.
XYZ Laboratories
2-5 Matsukage, Takamiya
Atsugi, Kanagawa 253-4321
Japan

上記の住所の書式をインラインの住所に変えるとき，以下の例のように，各行の末尾にコンマをつければよい。

- ABC Corp.**,** XYZ Laboratories**,** 2-5 Matsukage, Takamiya**,** Atsugi, Kanagawa 253-4321**,** Japan

Punctuation: Comma 2

5. 同じ動詞の繰り返しを避けるために，コンマを使う場合もある

- The MUX <u>has</u> 1402 elements; and the DEMUX, 1151.
  コンマは「has」を意味する。
- In air, <u>we obtained</u> single-phase Mn_2O_3; and in hydrogen, single-phase MnO.
  2番目のコンマは「we obtained」を意味する。

PRACTICE: 次の文にコンマを入れよ。

1. However the photovoltaic parameters are very low. For example the fill factor is only 44% and the conversion efficiency is just 0.07%.
2. It is difficult to estimate the spring constant κ and the damping factor γ.
3. Mori University Building 65 W-Wing 804C 5-2-9 Okubo Shinjuku Tokyo 170-8432 Japan
4. The wavelength was 203 nm; the intensity ~110 mJ/cm^2/pulse; and the repetition rate 16 pulses/s.
5. One source is the variation in the neutral current which is caused for example by an unbalanced load.
6. One plate is soldered to Package 1; and the other to Package 2.
7. The junction temperature T_j was about 225°C.
8. This study employed the LQR optimal control method (see for example [14]).
9. The transmitter consumes 1.2 W; and the receiver 0.75 W.
10. For example if the key "2222" is input the flag for address "2222" is read.
11. To tune the output frequency we can change either the reference frequency f_{comb} or the multiplication order N.
12. Linkers Inc. 6F Sato Building 4-7-6 Ginza Chuo Tokyo 106-0328 Japan
13. When the emitter-base voltage V_{BE} was low the base current I_B increased significantly with time.

Section 5

Unnecessary Words 2

「Is used」という動詞は余分になっている場合が少なくない。

次のペアとなる文を吟味しよう。
- A computer <u>is used to control</u> the movable mirrors.
- A computer <u>controls</u> the movable mirrors.

2番目の構文の方がシンプルかつ直接であり、はるかによい。

PRACTICE: 不必要な「is used to」を削除せよ。

1. The first test <u>was used to</u> clarify the degradation modes.

   _____

2. Optical fibers <u>are used to</u> transport light to and from the probe.

   _____

3. A Teflon lens <u>was used to</u> focus the signal on the detector.

   _____

4. This terraced structure <u>is used to</u> improve the heat dissipation of the chip.

   _____

5. An electro-optic probe <u>was used to</u> detect reflected signals.

   _____

6. The tungsten <u>is used to</u> reduce the resistance of the electrodes.

   _____

7. Etching with HF vapor <u>is used to</u> remove the SiO_2 sacrificial layer.

   _____

8. These chemical agents <u>are used to</u> improve the adhesion of polymer to a substrate.

   _____

Prepositions 5

空欄を適切な前置詞で埋めよ。必要がなければ、×で埋めよ。

REVIEW

a. X is composed _____ Y and Z.

b. X is equal _____ Y.

c. X corresponds _____ Y.

d. Result X agrees _____ result Y.

e. X is sandwiched _____ Y and Z.

f. X is related _____ Y.

g. X takes Y _____ account.

h. a change _____ the voltage

i. X is in good agreement _____ Y.

j. X is connected _____ Y.

CHECK YOUR KNOWLEDGE

1. X is associated _____ Y.

2. X gives rise _____ Y.

3. X is superior _____ Y.

4. As a result _____ X,.....

5. a decrease _____ the rate

6. There is a great deal of data _____ Y.

7. X is suited _____ Y.

8. X is characteristic _____ Y.

9. fluctuations _____ the temperature

10. _____ contrast, ...

Section 6

know vs. find out

approach to   key to

then

a/an   vs.   one of (the)

most   vs.   most of (the)

none, one, some, most, all

Meaningless –ing

issue vs. problem

obvious

so-called

optics is vs. optics are

therefore vs. so

problem with/of

Punctuation: Semicolon (;)

Style: (Fig. 3)

Prepositions 6

know vs. find out

以下の図と文を吟味しよう。

ACTION
He is putting on his socks now.
今靴下を履いているところである。

STATE
He is wearing socks.
靴下を履いている。

この例では、英語は行動と状態に**別の言葉**を使うことに対して、日本語は**同じ言葉**の違う形を使っている。それは英語と日本語の面白い違いである。他の例もある。

put on glasses	**wear** glasses
put on clothes	**wear** clothes
get sick（病気になる）	**be** sick（病気になっている）
die（死ぬ）	**be dead**（死んでいる）

このような表現を訳すには特に注意しなければならない。重要な例は以下に示す。

What is 7 x 8?

TO FIND OUT...
それを知るために…

I KNOW!
それを知っている！

POINT 1:「知る」という単語は、情報を得る行動を意味する場合、「**find out**」と訳す。

POINT 2:「知る」という単語は、情報を持っている状態を意味する場合、「**know**」と訳す。

Typical Mistakes

- To (~~know~~) **find out** whether or not the structure contains SiO_2, we analyzed the interior.
- To obtain good control of the device characteristics, we must (~~know~~) **find out/gain a thorough understanding of** how the oxidation proceeds.
- We can (~~know~~) **find out/determine** which LSI chips are good by measuring their standby current at a low temperature.

Good Examples

- **If we know** the surface potential, we can derive the thickness.
- We **need to know** more about how these resists work.
- We **know** that the defects contain oxygen.
- **As you know**, it is important to reduce the power consumption of LSIs.

Section 6

PRACTICE: 空欄を「know」か「find out」で埋めよ。

1. We measured the change in weight of a calcium electrode to _____ whether or not electrochemical dissolution occurs.
2. To fully understand self-heating-related phenomena, we need to _____ the device temperature.
3. So, it is very important to _____ how these services are actually used if we hope to reduce their environmental load.
4. We need to _____ what concentration of surfactant makes water miscible in hexane.
5. We also ran tests in Hokkaido to _____ how the battery pack would perform in intense cold.
6. Plasma containing a large amount of fluoride gas does not etch Ta. To _____ why, we added a small amount of fluoride gas to Cl plasma and investigated its influence.
7. If we _____ the width and length, we can estimate the height using this relationship.
8. To _____ what is happening during the incubation period before deposition begins, we analyzed the surface of deposited film.
9. We _____ that AsH$_3$ reacts with Group-III sources.
10. Activation energy is a useful parameter for characterizing reactions. So, we need to _____ what it is to clarify the degradation process.

approach to　　key to

POINT: 「Approach」と「key」の後ろの「**to**」は前置詞である。前置詞の目的語は名詞か動名詞であり，動詞ではない。

one approach	to improving…	○
the key	to the improvement of…	○
	~~to improve~~	✗

- A new **approach to** (~~improve~~) **improving** resist patterns has been developed.
- Our **approach to** (~~shorten~~) **shortening** the switching time is explained below.
- Another **approach to** (~~prevent~~) **preventing** oscillations was devised.

- Sealing with polyimide is the **key to** (~~improve~~) **improving** the yield.
- These technologies are the **keys to** (~~make~~) **making** ultralow-power LSIs.
- This is the **key to** (~~reduce~~) **reducing** roughness.

then

POINT: 技術英語では「then」はおもに以下の三つの意味に使われている。

行動の順番: First ... Next ... Then ...

- **First**, the sample was rinsed. **Next**, it was put in the dryer chamber. **Then**, the chamber was heated.
- The voltage was scanned up to −2.1 V, and **then** back down.
- Each bit is reshaped, and **then** stored temporarily in an output buffer.

条件または仮定の論理的な結果

条件または仮定は, if-節の形でまたは「assume」や「suppose」という言葉を使って表現される。If-節が短い場合には,「then」をよく省略する。

- **If** there is a fault here, **(then)** this circuit will never be reset.
- **If** the width of the opening is much smaller than its depth, **(then)** it is difficult to fill the opening completely with metal.
- **Assume** that $\Delta M(q)$, $\Delta H(q)$, and τ_{ext} are all zero. **Then**, combining (5) and (7) yields
$$\ddot{e} + K_D \dot{e} + K_P e = 0. \tag{8}$$

「Let」や「set」の後ろの論理的な結果

「Let」や「set」を使って値を変数に代入した後で, その代入することによる論理的な結果を示すために「then」を使うことができる。

- **Let** the reference input be
$$r(t) = \sin\pi t + 0.5\sin 2\pi t + 0.5\sin 3\pi t. \tag{38}$$
Then, the repetition period is
$$T = 2 \text{ s}. \tag{39}$$

Typical Mistakes

then ≠ だから, それで

- A larger area means more expensive chips. (~~Then~~) **So**, we have to limit this increase in area when we apply our technique.
- This increases the refractive index, (~~then~~) so the wavelength becomes longer.
- The slopes of these lines are almost the same. (~~Then~~) **Thus**, the failure modes are probably the same.

「Then」は等位接続詞ではない。

- ✗ *The electric field in the absorption layer becomes stronger***,** *<u>then</u> it accelerates the photoelectrons in that layer.*
- ○ The electric field in the absorption layer becomes stronger**.** **As a result,** it accelerates the photoelectrons in that layer.

a/an vs. one of (the)

✗ *A tiger is <u>one of the animals</u>.*
✗ *A tiger is <u>one of the kinds of animals.</u>*

以下の説明でわかるように，animals も kinds of animals も特定のグループではないから，「one of」は間違いである。

一般的な場合

POINT: 通常,「…の一つ」は「**a**」か「**an**」に訳すのが一番よい。

○ I ate **an** apple.

○ A tiger is **an** animal.
○ A tiger is **a** kind of animal.

特定のグループ

POINT: 特定のグループの中の一つのものを述べるとき「**one of**」を使う。

○ I ate **one of** <u>the apples in the box</u>.

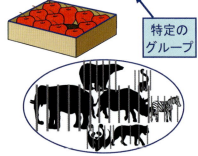

特定のグループ

○ **One of** <u>the animals in Ueno Zoo</u> escaped.
○ A tiger is **one of** <u>the most dangerous animals</u>.

Typical Mistakes

one of the a/an

- Area bump technology is (~~one of the solutions~~) **a** solution.
- V_p is (~~one of the crucial parameters~~) **a** crucial parameter.

✗ *Making wires narrower is <u>one of the</u> effective ways to reduce…*
○ Making wires narrower is **an** effective way to reduce…
○ Making wires narrower is **one of the most effective ways** to reduce…

✗ *Differential drive is <u>one of the</u> common designs in bipolar circuits.*
○ Differential drive is **a** common design in bipolar circuits.
○ Differential drive is **one of the most common designs** in bipolar circuits.

most vs. most of (the)

一般的な場合	特定のグループ（可算名詞） 特定のもの（不可算名詞）
POINT: ある種の物事の**すべて**について一般的に述べるとき「**most**」を使う。	**POINT:** ある**特定のグループか物事の一部**について述べるとき「**most of**」を使う。

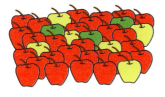

- **Most** <u>apples</u> are red.

- **Most of** <u>the apples in the box</u> are yellow. 　　　特定のグループ

- **Most** <u>oil</u> is deep underground.

- **Most of** <u>the oil imported into Japan</u> comes from the Middle East. 　　　特定の物

Typical Mistakes

- ✗ *Most of optical receivers have been designed for continuous signals.*
- ○ **Most** <u>optical receivers</u> have been designed for continuous signals.

- ✗ *Most of the semiconductor mode-locked lasers have a linear cavity.*
- ○ **Most** <u>semiconductor mode-locked lasers</u> have a linear cavity.

- ✗ *Most of peripheral functions of the power supply card are handled by a microcontroller.*
- ○ **Most of** <u>the peripheral functions of the power supply card</u> are handled by a microcontroller.

- ✗ *Most of flammable material in a Li-ion battery can be eliminated by…*
- ○ **Most of** <u>the flammable material in a Li-ion battery</u> can be eliminated by…

none, one, some, most, all

いくつかの単語は，most—most ofと同じパターンに従う。

一般的な場合

POINT: ある種の物事の**すべて**について一般的に述べるとき，以下の表現を使う。

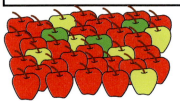

- **No** <u>apples</u> are blue.

- **An** <u>apple</u> a day keeps the doctor away. (ことわざ：1日1個のリンゴで医者要らず)
- **Some** <u>apples</u> are yellow.

- **Most** <u>apples</u> are red.

- **All** <u>apples</u> grow on trees.

特定のグループやもの

POINT: ある**特定のグループか物事の一部**について述べるとき，以下の表現を使う。

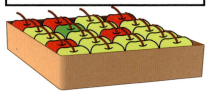

- **None of** the apples in the box is blue.
- **One of** the apples in the box is green.

- **Some of** the apples in the box are red.
- **Most of** the apples in the box are yellow.
- **All (of)** the apples in the box are delicious.

NOTE: 他の同じパターンに従う言葉は以下のものがある。
(a) few　　any　　many　　much　　several

PRACTICE: 例にならって，括弧に囲まれている言葉を文に入れよ。
- Ex. 1: (*a/an/one*) Thermal frequency drift is __an__ obstacle(s̶) to fast switching.
- Ex. 2: (*a/an/one*) __One of the__ sample gas(es) we selected was water vapor.
- Ex. 3: (*some*) __Some__ bits are corrupted during transmission.
- Ex. 4: (*some*) __Some of the__ integrated devices that were fabricated were measured.

1. (*some*) Table 1 lists _____ specifications of the new modulator.
2. (*a/an/one*) Phase noise is _____ big problem(s).

3. (*many*) The process is complex, and _____ process factors are coupled.
4. (*a/an/one*) _____ variable(s) in the formula is maximum heart rate.
5. (*many*) There are _____ concrete structures in the world.
6. (*all*) _____ components are commercially available.
7. (*some*) Section 5 presents _____ experimental results.
8. (*many*) This remote-driving scheme avoids _____ problems associated with contact-based schemes.
9. (*most*) Since no surface waves are generated, _____ power goes into the waveguide.
10. (*all*) _____ telecommunications devices require a low loss and a large dynamic range.
11. (*most*) Terahertz waves can penetrate _____ non-metallic materials.
12. (*some*) Millimeter waves have _____ interesting applications.
13. (*all*) _____ measurements were carried out at a temperature of 40°C.
14. (*most*) _____ high-speed digital-to-analog converters are based on this architecture.
15. (*all*) _____ possible pairs of the items are formed, and a search is carried out on each pair.
16. (*some*) _____ unit cells are arrayed two dimensionally.
17. (*a/an/one*) N_2O has _____ absorption line(s) at 16.6 μm.
18. (*a/an/one*) PECST is _____ Japanese national project(s).
19. (*most*) _____ light injected into the Ge is absorbed.
20. (*a/an/one*) A mobility of 22,000 cm²/Vs is _____ highest value(s) ever reported.
21. (*some*) This new device overcomes _____ limitations of conventional devices.
22. (*most*) _____ frequency components are below 10 Hz.
23. (*a/an/one*) _____ electrode(s) of our battery consists of MnO_2.
24. (*many*) Assumption 2 holds in _____ real-world systems.

Meaningless –ing

POINT: 現在分詞 (-ing) 句の主語はメイン動詞の主語とは同じでなければならない。主語が同一でない間違いは懸垂分詞（dangling participle）という。英語を母語としている人でもこの間違いをよく犯す。

✗ *After eating* lunch, *an e-mail* arrived.

「Eating」の主語が「an e-mail」となっているので，この文は次のような意味になってしまう。

✗ *After an e-mail ate lunch, it arrived.*

正しい文は以下のようになる。
- ○ After eating lunch, I received an e-mail.
- ○ After lunch, an e-mail arrived.

技術英語において，この問題は大抵**受動態**または**名詞**を用いて簡単に解決できるが，文を完全に書き直さなければならないこともある。

✗ *After depositing* SiO_2, *an area for the gate is opened by etching.*
○ After SiO_2 **is deposited**, an area for the gate is opened. （受動態）
○ After the **deposition** of SiO_2, an area for the gate is opened. （名詞）

PRACTICE: 次の文を直せ。

1. After shortening the annealing time, the blue shift disappeared.
   _____

2. Combining these functions, various types of processing can be performed.
   _____

3. Using these structures, a high mobility was obtained.
   _____

4. Analyzing the results, there are two patterns.
   _____

5. Using this property, interesting imaging applications have recently been developed.
   _____

例外： 以下の熟語はこの限りではない。

according to	allowing for	assuming	concerning
considering	depending on	excluding	generally speaking
owing to	regarding	strictly speaking	taking ___ into account

issue vs. problem

POINT: 「Issue」は解決または克服するべきものではなく，考慮すべき（頭におくべき）事柄である。

discuss

Issue は人々が論じている重要なトピックである。技術論文でよく論じるトピックまたは問題点に言及するとき，「**issue**」を用いる。

Issueの例

cost

health & safety

yield（歩留まり）

wavelength stability

- Suppressing the leakage current is **the main issue** in the fabrication of ultralow-voltage LSIs.
- In this type of network, wavelength management is **the key issue**.
- Power consumption is becoming **a big issue** as network capacity increases.
- The aging of society necessitates serious consideration of the **issue** of nursing care for the elderly.

solve

Problem とは，解決を必要とする事柄を指す。通常 **problem** は悪い事柄である。

Problemの例

the high cost of the device

the use of toxic（有毒）chemicals for fabrication

low yield

how to reduce temperature fluctuations

- **The remaining problem is** the large coupling loss between a waveguide and a fiber.
- **One problem with** these switches **is** that the performance of the driver amplifiers limits the operating speed.
- However, **there is a problem with** the conventional method.
- **Another problem is how to** stabilize the wavelength when the wavelength is switched.

何かをどのようにするかは通常problemである。

Section 6

> **PRACTICE:** 「Issue」か「problem」を次の空欄に入れよ。

1. Dirt on the sensor surface is a serious _____ for fingerprint sensors.

2. One _____ is the stability of the transmission wavelength.

3. Cost performance is always a(n) _____ because the construction of multichip modules involves many costly techniques.

4. One _____ with a dielectric filter is that the transmission wavelength depends on temperature.

5. To solve the phase fluctuation _____, we developed a planar lightwave circuit.

6. User authentication (ユーザー認証) to prevent the unauthorized use of equipment has become an important _____.

7. The mounting of optical devices, such as laser diodes, is an important _____.

8. The final section discusses a(n) _____ that will soon become important in the photonic measurement of microwaves: integration and packaging technology.

9. The use of stabilization circuits is a good way to solve this _____.

10. The design _____ is how to achieve high-speed gain control.

11. The main _____ in extreme-ultraviolet lithography (EUVL) is the production of defect-free masks.

obvious

POINT: 「Obvious」はしばしば読者に失礼な印象を与える。

DEFINITION: 何かが obvious とは，それはきわめて容易に理解できることである。すなわち，馬鹿な人でも理解できるほど明白である。

~~It is obvious (that)~~ ➡ *[Nothing]*, **It is clear that**

✗ *It is obvious that* the oxygen-deficient atmosphere produced a polycrystalline film.
○ The oxygen-deficient atmosphere produced a polycrystalline film.
○ **It is clear that** the oxygen-deficient atmosphere produced a polycrystalline film.

- To obtain a large extinction ratio, (~~it is obvious that~~) the light beams from the two arms of the modulator must have the same intensity.
- (~~It is obvious~~) **It is clear** from the graph that the pitch of interconnections is limited by the current density of via plugs.
- (~~It is obvious that~~) **Notice that** the multiplication factor is greater at long wavelengths than at short wavelengths.

~~obvious~~ ➡ **clear, pronounced, prominent, observable, etc.**

- In this spectrum, the Si 2p peak is (~~obvious~~) **pronounced/large/prominent/high**.
- Every data pulse is demultiplexed without any (~~obvious~~) **observable/apparent** crosstalk.

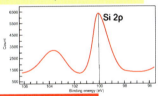

「Obvious」という言葉を使うとき，特に注意しよう。この単語は技術英語ではあまり使われていない。

so-called

DEFINITION: 「So-called」は，間違っているか，誤解を招くおそれがある言葉を導入するのによく使われる。

Good Examples

- Sai Baba is a so-called "guru" with many followers.
  Guru は精神的リーダーであるが，Sai Babaはただのマジシャンであり，宗教指導者ではない。

技術英語では，「so-called」はほとんど使わない。

Section 6

~~so-called~~ → *[Nothing]*, what is called, what (they) call

Typical Mistakes

✗ *This configuration produces a <u>so-called</u> off-axis Fourier hologram.*
○ This configuration produces an off-axis Fourier hologram.
○ This configuration produces **what is called** an off-axis Fourier hologram.
　読者に，ある専門用語が馴染みがないと思う場合，それを紹介するのに「what is called」を使う。

✗ *The MIT group suggested the <u>so-called</u> barrier-induced hole pile-up model.*
○ The MIT group suggested the barrier-induced hole pile-up model.
○ The MIT group suggested **what they call** the barrier-induced hole pile-up model.
　ある専門用語は誰が作ったかを示すのに「what (they) call」を使う。

✗ *Yamaguchi et al. developed the <u>so-called</u> chip-size-cavity package.*
○ Yamaguchi et al. developed the chip-size-cavity package.
○ Yamaguchi et al. developed **what they call** a chip-size-cavity package.

optics is vs. optics are

POINT: 「Optics」,「dynamics」,「electronics」,「statistics」などの単語は，科学や数学の<u>分野</u>を意味する場合，<u>単数形</u>であり，<u>特定の</u>システム，構成要素，または数値を意味する場合，<u>複数形</u>である。

optics
（単数形） <u>Optics is</u> the study of light.
（複数形） The light collection <u>optics provide</u> efficient optical coupling.
（複数形） <u>These</u> free-space <u>optics transmit</u> data at a rate of 10 Gb/s.

electronics
（単数形） <u>Electronics deals</u> with the flow and control of electrons.
（複数形） In an LED lamp, the <u>electronics limit</u> the lifetime to 20,000 hours, even though the LED itself will last 100,000 hours.

statistics
（単数形） <u>Statistics provides</u> tools for analyzing data.
（複数形） Previously collected <u>statistics were</u> used to determine the minimum and maximum loads.

dynamics
（単数形） <u>Dynamics is</u> a branch of mechanics.
（複数形） The nonlinear <u>dynamics</u> of the system <u>are</u> difficult to model.
（複数形） The <u>spin dynamics</u> during electron diffusion <u>were</u> measured.

therefore vs. so

A = B and B = C.
Therefore,
A = C.

The computer is broken. So, we cannot use it.

DEFINITION:「Therefore」を使って論理的な結果や結論を導く。

DEFINITION:「So」を使って，その前に説明したことの成り行きを述べる。

 ➡ so, thus, as a result, etc.

Typical Mistakes

- A networked control system (NCS) has many advantages: quick and easy maintenance, low cost, great flexibility, etc. (~~Therefore~~) **So**, NCSs are increasingly being used for industrial control in a variety of fields.

- The phosphorus distribution of WSiN films varies with thickness. (~~Therefore~~) **So**, we investigated the dependence of thickness on the nitridation conditions.

- These systems use fixed-wavelength lasers, and component vendors must now stock many lasers for each wavelength in case of trouble. (~~Therefore~~) **So**, inventory costs can be considerably reduced if a widely tunable laser is used as a replacement part.

Good Examples

反応を与える可能性のある物質は，(1)酸素，(2)オゾン，のいずれかとわかっているが，実験の結果，(1)は否定された。
- These results eliminate oxygen as a possible reactant. **Therefore,** the initial stage of the reaction must involve ozone

- The maximum radiated power is 140 µW for each antenna element. **Therefore**, the maximum radiated power for the antenna array (9 elements) is about 1 mW.
1 mWは計算した結果であり，計算は論理演算である。

「Therefore」は等位接続詞ではない。

✗ *The transmission distance of this wireless link is about 1 km, therefore its applications are limited.*

○ *The transmission distance of this wireless link is about 1 km, so its applications are limited.*

problem with/of

> **POINT:** ある物事におかしい所，不十分な面，または欠点がある（すなわち，ある物事に問題がある）場合は，通常「**problem with**」を使って説明する。

problem with　　Zには問題がある。　　There is a **problem with Z**.
　　　　　　　　　Zの問題は__である。The **problem with Z** is (that) __.

- The **problem with** the car **is** the brakes.
  車は問題ではなく，車に問題がある。
- The **problem with** the car **is that** the brakes do not function properly.
- The **problem with** the car can be solved by replacing the brake pads.

Typical Mistakes

- The problem (~~of~~) **with** this device is the cost of assembly.
- The problems (~~of~~) **with** this method are the large line width and poor stability.

problem of　　Zという問題　　　　the **problem of Z**

- The **problem of** the brakes can be solved by replacing the brake pads.
  ブレーキ自体は問題である。
- This leaves the **problem of** <u>how to</u> control the phase error in the delay lines.
  何かをどのようにするかは通常 problem である。
- This type of laser suffers from the **problem of** mode hopping.

> **PRACTICE:**「Of」か「with」で空欄を埋めよ。

1. One problem _____ current EAMs is the driving voltage, V_{pp}.
2. The problem _____ previous approaches is discussed below.
3. This section concerns the problem _____ wavelength deviations due to temperature fluctuations.
4. One problem _____ a dielectric filter is that the transmission wavelength depends on temperature.
5. A serious problem _____ Si wire waveguides is the coupling to a fiber.
6. Once the crucial problem _____ coupling loss was solved, our attention turned to the fabrication of functional devices.
7. Both types of devices suffer from the problem _____ crosstalk due to the bonding wires.

Semicolon (;)

1. リスト中の長い項目またはコンマを含む項目の分割

- The equivalent circuit (Fig. 4) has three parameters: the resistance of a bump, R_B; the inductance of a bump, L_B; and the capacitance between the signal pads and the ground, C_P.

- It was implemented to have three features: easy, stable operation; easy development and debugging; and simple maintenance.

- Possible applications of terahertz waves include quality inspection for industrial and agricultural products; the detection of concealed weapons, explosive substances, and controlled drugs at security gates; medical diagnosis; and broadband wireless communication systems.

2. コンマを含む主節の分割

- When SE is low, the data comes from normal D input; and when it is high, the data comes from scan-in input.

- The deposition rate depends little on hydrogen gas pressure; but the deposition constant, K, is probably influenced by equipment-related factors, such as the method of heating and the pumping speed.

- The amplitude of the resonance oscillations <u>ranges</u> from 1 to 100 μm; the resonance frequency, from 100 Hz to 1 GHz; and the quality factor, from 10^3 to 10^6.

  この文で，コンマは「ranges」という動詞を代表している。このパターンは同じ言葉の繰り返しを避けるために使用される。

3. 密接に関係する二つの節の分割

- It has been observed that, above a value of 1, it becomes difficult or impossible to simultaneously print all feature types at the correct size; this can be used as a rule of thumb to determine the resolution limits of the different processes.
  この場合，節を二つの文に分けることができる。
  > It has been observed that, above a value of 1, it becomes difficult or impossible to simultaneously print all feature types at the correct size. This can be used as a rule of thumb to determine the resolution limits of the different processes.

Section 6

> **DANGER:** 数値または定義を導入するために，セミコロンではなく，コロンまたは等号を使用すべきである。
>
> ✗ *A fixed temperature (T_g; 350°C) was used for growth.*
> ○ A fixed temperature (T_g: 350°C) was used for growth.
>
> m; ~~mass of fine stage~~   → m: mass of fine stage
> M; ~~mass of coarse stage~~ M: mass of coarse stage

PRACTICE: 次の文中の不適切なコンマをセミコロンに変えよ。

1. The open circles are for the pile-up model, and as you can see, the agreement is excellent.

2. This figure illustrates a number of key characteristics of the kink: The kink in I_D occurs approximately at a constant V_{DG} of 1.2 V, the size of the kink appears to increase with increasing V_{GS}, and the onset of the kink coincides with the appearance of I_{SG} and with a prominent rise in E_G, presumably due to hole collection by the gate.

3. Below the line, diffusion is dominant, and above, drift is dominant.

4. Unlike digital LSIs, most of the area of MMICs is occupied by passive elements, such as transmission lines, inductors, and capacitors, and reducing their size is the best way to miniaturize MMICs.

5. Circuit extraction is an important step in VLSI circuit design verification, it provides the link between the physical design and its verification phases.

6. As shown in Fig. 8, when the phase is 160°, the best-focus position does not shift at all, and the depth of focus is as wide as that without spherical aberration.

7. The data timing is ideally centered at zero and tracks the bit length, it is ideally ±0.75 ns at a bit length of 1.5 ns, and ±1.5 ns at a bit length of 3.0 ns.

(Fig. 3)

「Figure 3 shows」,「as shown in Fig. 3」のような表現を通常「**(Fig. 3)**」というように非常に簡単に表現できる。これによって、アイデア表現の流れがスムーズになり、文章はもっとダイナミックになる。

Ex. 1: The XYZ <u>circuit diagram is shown in Fig. 13</u>. The circuit consists mainly of an inductor and four diodes.
　⇨ The XYZ circuit **(Fig. 13)** consists mainly of an inductor and four diodes.

Ex. 2: <u>Figure 5 shows</u> the optical signal spectrum. A dual-peak spectrum with a bandwidth of 75 GHz <u>was obtained</u>.
　⇨ The optical signal spectrum **(Fig. 5)** has dual peaks and a bandwidth of 75 GHz.

Ex. 3: An experiment was performed using discrete ICs and ceramic capacitors. The measurement system <u>is shown in Fig. 2</u>.
　⇨ Discrete ICs and ceramic capacitors were used to build a measurement system **(Fig. 2)**.

PRACTICE: 以下の文を (Fig. X) のスタイルに変えよ。

1. Figure 2 shows the situation in which the receiver is near a conductive wall. The wall changes the pattern of the quasi-electrostatic field.

　⇨ If the receiver is _____.

2. A TORA model is shown in Fig. 6. It consists of a cart and an eccentric rotational proof mass.

　⇨ A TORA _____.

3. Figure 5 shows the simulation results. As can be seen in the figure, the settling time was less than 13 s.

　⇨ The simulation results _____.

4. The control system is shown in Fig. 1. It consists of four parts.

　⇨ _____.

5. The relationship between β and J_1 is shown in Fig. 5. It is clear that J_1 is small when $0.25 < \beta < 0.55$.

　⇨ _____.

6. Simulation results for the two parameter sets are shown in Fig. 7. Clearly, the control performance is better for Set B than for Set A.

　⇨ _____.

7. Figure 5 shows the topology of the model. It has 13 nodes.

　⇨ _____.

Section 6

Prepositions 6

空欄を適切な前置詞で埋めよ。必要がなければ，×で埋めよ。

REVIEW

a. X is comparable _____ Y.

b. X is similar _____ Y.

c. Sato et al. mention _____ two problems ...

d. X gives rise _____ Y.

e. X is characteristic _____ Y.

f. X is associated _____ Y.

g. X is suitable _____ Y.

h. X is suited _____ Y.

i. X is added _____ Y.

j. X affects _____ Y.

CHECK YOUR KNOWLEDGE

1. X arises _____ Y.

2. the variation _____ the width

3. X is made up _____ Y and Z.

4. X is injected _____ Y.

5. X is incorporated _____ Y.

6. This paper deals _____ X.

7. X compensates _____ Y.

8. X is consistent _____ Y.
   この表現を「consist of」と混同しないこと。

9. X takes Y _____ consideration.

10. The limit _____ X...

Section 7

measured vs. measurement

contribute to

Bad Passives

because vs. since

composition vs. content

another vs. the other

maintain vs. remain

improve

saturate

not A or B

be consistent with

recover vs. restore

monotonous vs. monotonic

Punctuation: Parentheses

Style: Larger For A Than For B

Prepositions 7

measured vs. measurement

POINT 1: 「**Measured wavelength**」も「**measurement wavelength**」も正しいが，意味が違う。

- The **measured wavelength** was 400 nm.
  波長の測定結果は 400 nm である。
- The **measurment wavelength** was 400 nm.
  測定に用いる波長は 400 nm である。

POINT 2: 「**Results**」という単語の使い方に常に注意する必要がある。「**Measured**」のような過去分詞は前につけてはいけない。その理由として，測定の結果を測定できないからである。そのため，「**measured results**」は間違いである。

We **measured the characteristics** and obtained some **results**.

⬇ ⬇

The characteristics **were measured**.　　　the results **of the measurement**

⬇ ⬇

measured characteristics　　　**measurement results**

Similarly,　　　Similarly,
　calculated characteristics　　　**calculation results**
　simulated characteristics　　　**simulation results**

- From the **simulated waveforms** of a 32-bit BLC adder (Fig. 6), we see...
- Table 3 compares the **simulated and measured power consumption**.
- The threshold current density was estimated based on the **calculated carrier distribution**.

- The **measurement results** on fast tuning (Fig. 2) show that...
- The **measured frequency characteristics** agree well with the **calculation results**.
- The **simulation and measurement results** were compared to verify...

contribute to

POINT: 「**Contribute**」の後ろの「**to**」は前置詞である。前置詞の目的語は動詞ではなく，名詞または動名詞である。

contribute ⎰ to the improvement of…　　○
　　　　　 ⎨ to improving…　　　　　　　○
　　　　　 ⎱ ~~to improve…~~　　　　　　×

- This **contributes to** (~~improve~~) **improving** the reliability.
- Our goal is to **contribute to** (~~create~~) **the creation** of an environmentally friendly society.
- This **contributes to** (~~reduce~~) **the reduction** of CO_2 emissions.

Bad Passives

> **POINT:**「**Occur**」,「**remain**」,「**appear**」と「**disappear**」は自動詞であり，受動態はない。「**Originate**」は他動詞でも自動詞でもあるが，自動詞の方がよく使われる。

occur （自動詞）

- ✗ *Anode degradation due to carbon deposition is occurred.*
- ○ Anode degradation due to carbon deposition **occurs**.
  場合によっては，名詞を動詞に変えることにより「be occurred」という部分が省略できる。
  degradation → degrade
- ○ The anode **degrades** due to carbon deposition.
- ○ Carbon deposition **causes** the anode **to degrade**.

- ✗ *The electrolyte decomposition is occurred continuously.*
- ○ The electrolyte continuously **decomposes**.

remain （自動詞）

- This method expands only narrow valleys, so narrow ridges (~~are remained~~) **remain**.

appear/disappear （自動詞）

- The degradation (~~is only appeared~~) **only appears** during impact stress.

originate (in, from) （自動詞）

- The incubation period (~~is originated from~~) **originates from** the reduction in the amount of adsorbed oxygen.
- Our technique compensates for the skew (~~originated in~~) **originating in** the output cables.

「Originate」を他動詞として使う場合，主語は人であることに注意しよう。例えば
- Einstein **originated** the theory of relativity.

PRACTICE: 間違っている受動態の動詞を直せ。

1. The variation **is originated in** the fabrication process.
   _____.

2. A thin layer of SiO2 **was remained** on top.
   _____.

3. The GaAs layer **may be disappeared** during etching if…
   _____.

4. Limit-cycle oscillations **is occurred**.
   _____.

Section 7

5. It is impossible to avoid the bandwidth limitation **originated from** the carrier response of the semiconductor.
   _____.

6. When a bit error **is occurred**, the device requests that the data be resent.
   _____.

7. The data in the selected memory cells **are appeared** on the bit lines.
   _____.

8. This high value suggests that some hydrogen **is still remained**.
   _____.

because vs. since

since

GUIDELINE: 「Since」で始まる原因の節は, 結果としての主節を強調するために, 通常主節の前におく。

- **Since** the clamping force is greater than the buckling force, no adhesive is necessary.
- **Since** the Arrhenius plot is roughly linear, it can be concluded that the drop in capacity is due to a chemical reaction.

because

GUIDELINE: 「Because」で始まる節は, 原因を強調するため, 通常主節の後ろにおく。

- For mobile equipment, power efficiency is the top priority **because** it determines battery life.
- This laser is easy to tune **because** it needs only one tuning current.

DANGER: 「Because」で始まる節自身は完全な文ではなく, ただの副詞節である。

✗ The coefficients of thermal expansion of the chip and the film must be almost the same. _Because a large difference between them reduces the adhesion._
○ The coefficients of thermal expansion of the chip and the film must be almost the same **because** a large difference between them reduces the adhesion.

しかし, 質問の答えとして, 「because」で始まる節を単独に使ってよい。

○ **Why** do we integrate IC chips on a PLC substrate? **Because** it results in smaller, cheaper devices.

composition vs. content

composition

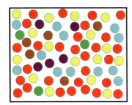

「**Composition**」はある材料が構成されている物質の**すべて**を指す。

- The **composition** of air is
  - N_2 78.08%
  - O_2 20.95%
  - Ar   0.93%
  - CO_2 0.03% (avg.)
  - etc.
- The absorption spectrum of a molecule depends on its **composition** and structure.

content

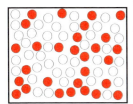

「**Content**」はある材料が含まれている**一つ**の物質の割合を指す。

- The nitrogen **content** of air is 78.08%.
- Secondary ion mass spectrometry (SIMS) revealed the Mn **content** to be 1%.
- Imported iron ore generally has a low sulfur **content**.

Typical Mistakes

the X content ~~in~~ Y the X content **of** Y

- The dielectric constant increases as the moisture **content** (~~in~~) **of** the film increases.
  水分はフィルムの中にある。しかし，**水分の含有量**は，濃度，熱伝導率，厚さなどと同じように**フィルムの特性**であり，フィルムの中の物ではない。
- The dependence of the In **content** (~~in~~) **of** quantum wells on the TMIn **content** (~~in~~) **of** the vapor was examined.

PRACTICE: 空欄を「composition」か「content」で埋めよ。

1. The indium _____ and the thickness were adjusted.
2. The chemical _____ of the film was analyzed.
3. Figure legend: Etching rate vs. CF_4 _____
4. The alcohol _____ of beer is printed on the label.
5. The shape of the domains depends on the _____ of the polymer.

another vs. the other

another

DEFINITION: 一つのものに言及した後，同種類の別のものに言及する場合，「another」を使う。

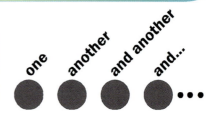

- One big advantage is that we do not need a phase control region. **Another** is easy stabilization of the wavelength.
- For identification, the sensor LSI captures a fingerprint image. If the quality is not good enough, the parameters are changed and **another** image is captured.
- Figure 6 shows the results of accelerated aging tests to estimate the lifetime. In **another** test, the cells were charged intermittently ...

✗ the another   **NOT ENGLISH**

「Another = an other」なので，「the an other」と書くことはできない。

the other

DEFINITION: 特定のグループについて話をする場合，そのグループの一部について述べた後，残りの部分を述べるとき「the other(s)」を使う。

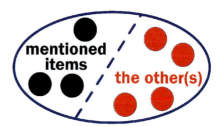

↙ 特定のグループ

- Two laser beams are fed into this loop. One is a reference, and **the other** is a pump laser.
- There are two ways to expand the exposure area. One is to move the mirror back and forth, and **the other** is to move the mask and wafer.
- There are various ways of depositing copper, such as sputtering, plating, and CVD. CVD has several big advantages over **the others**.
- When one of the four transistors is active, **the others** are inactive.

maintain vs. remain

remain　「**Remain**」は，変化しない**状態**が続くということだけを表している。

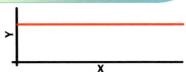

- Several fundamental problems **remain** unsolved.
- The chamber is not heated and **remains** at room temperature.
- This mechanism ensures that the output power **remains** constant.
- We need to find ways to help the elderly **remain** self-reliant.

maintain　「**Maintain**」はもっと**アクティブな**イメージがあり，同じ状態に保つための**行動**を指す。この単語が適切に使われているならば，「何が何を維持するか」または「何が何で維持されるか」という質問の回答になる。

- A capacitor in the loop filter **maintains** the control voltage of the oscillator.
- The pressure **is maintained** at 7.5 MPa with a pressure-control valve.
- Normal single-mode fiber **cannot maintain** the polarization of propagating light.
- This new-concept electric cart helps the elderly **maintain** their physical strength.
- The plug and the fiber guide are bonded together to **maintain** the connection after assembly.
- A phase-locked loop **maintains** the relative phase between the pump and signal.

「変化なし」ということ**だけ**を記したい場合，「**maintain**」は適切ではない。

Common Patterns

remain ＋ 形容詞

✗ *Good electroabsorption characteristics are maintained.*
○ The electroabsorption characteristics **remain good**.

remain the same

✗ *The center of the delay time distribution shifts to a larger value, but the critical delay is maintained.*
○ … but the critical delay **remains the same**.

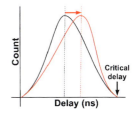

remain ＋ 数値

✗ *A current gain of over 20 is maintained.*
○ The current gain **remains over 20**.

Section 7

> **PRACTICE:** 「Maintain」を「remain」に変えよ。

1. A <u>constant</u> output power <u>is maintained</u>.

2. The extinction ratio <u>is maintained</u> over 13 dB.

3. The Ni and Ti contents <u>were maintained</u> during the ion-exchange process.

4. The optical quality of the output light <u>should be maintained</u>, even after thousands of circulations.

5. The <u>steep</u> energy profile around the edge <u>is maintained</u>.

6. The loss <u>maintained</u> a <u>sufficiently low</u> value.

improve

> **POINT:** 本質的に悪い物事は **improve** できない。

　薬を飲んで体調がよくなっていく場合，その薬は体調を improve した，または**体調が** improve したといえる。しかし，その薬は病気の程度を軽くしたので，薬は<u>病気を</u> improve したともいえるだろうか？答えは no である。ロス，損害，雑音，干渉，廃棄物のような**悪い物を** improve **しない**。通常そのものの量を減らしたり程度を低減する，またはそのものを完全には排除するというように表現する。

- ✗ *The interchannel <u>crosstalk was improved</u> to −22 dB.*
  クロストークはよくないものである。
- ○ The interchannel **crosstalk was reduced** to −22 dB.

- ✗ *Section 3 concerns the <u>improvement</u> of the optical <u>frequency drift</u>.*
  周波数ドリフトはよくないものである。
- ○ Section 3 concerns the **reduction** of the optical **frequency drift**.

- The insertion **loss** (~~improves~~) **decreases** as the substrate gets thinner.
- This technique (~~improves~~) **reduces** the mode **mismatch** at the tapered tip.
- The power **penalty** of the signal was significantly (~~improved~~) **reduced**.

saturate

POINT 1:「Saturate」は，数学の概念ではなく，物理学と化学の概念である。ただ変数があるレベルまで増加して，その後増加を停止したから，「saturate」を使うわけではない。

以下の例を吟味すれば，「**saturate**」の基本的なイメージがつかめられる。

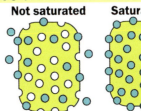

Ex. 1: スポンジに吸収できる限界まで水を含ませ，これ以上吸収できない状態となった場合，スポンジは水で飽和しているという。
- The sponge **is saturated** with water.

Ex. 2: The market **is saturated.** とは，消費能力の限界まで製品が市場に供給され，売上高が伸び悩んでいることである。

Ex. 3: 水の溶解できる限界まで食塩を水に入れると，飽和食塩水となる。
- …**saturated** salt solution…

Ex. 1 には，水が供給され，スポンジが限界までそれを受け入れる。
Ex. 2 には，製品が供給され，市場が限界までそれを受け入れる。
Ex. 3 には，塩が供給され，水が限界までそれを受け入れる。

上の例で示したように，「saturate」という概念には二つの要素がある。
　(1) 何かが供給されている。
　(2) 受け取る側の受け入れ量が増加して限界に達する。
「Saturate」の由来している意味はこれに基づいている。

例えば，増幅器に電圧を入力すると，それが増幅されて出力されるが，ある値以上の電圧が入力されると，出力電圧が飽和して一定値となる。

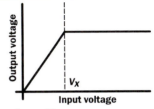

- The output voltage **saturates** at an input voltage of V_x.
- The output voltage **is saturated** at input voltages above V_x.

POINT 2:「Saturate」は「供給」と「受け入れ」の関係があるものにしか使えない。例えば，テレビの番組の視聴率が頭打ちになったことに，「saturated」は使えない。

Typical Mistakes

~~saturate~~ ➡ **level off**

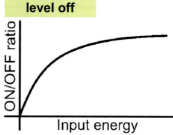

✗ *As the input energy increases, the ON/OFF ratio increases rapidly and then <u>saturates</u>.*
比率は「受け入れ量」ではない。

○ As the input energy increases, the ON/OFF ratio increases rapidly and **levels off**.

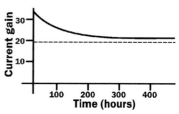

✗ *The current gain drops initially, and then the degradation appears to <u>saturate</u> at around 300 hours.*
低下の程度および電流ゲインはすべて減少しているため，「saturate」と正反対である。また，低下することは「受け入れ量」と関係ない。

○ The current gain drops initially, and then **levels off** at around 300 hours.

~~saturate~~ ➡ **reach a maximum**

✗ *The delay starts increasing at a voltage of −1.5 V, and <u>saturates</u> at a voltage of around −0.4 V.*
遅れは「受け入れ量」ではない。

○ The delay starts increasing at a voltage of −1.5 V and **reaches a maximum** at a voltage of around −0.4 V.

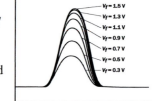

✗ *The height of the peak increases with V_f and <u>saturates</u> when $V_f = 1.5$ V.*
高さは「受け入れ量」ではない。

○ The height of the peak increases with V_f and **reaches a maximum** when $V_f = 1.5$ V.

Good Examples

○ The drain <u>current</u> increases with bias and **saturates** when the bias reaches 3 V.

○ The SH <u>power</u> **saturates** when the input power is increased.

○ The PC <u>market</u> **saturated** when the use of smart phones and other portable mobile devices became popular.

not A or B

POINT: AとBという二つのものを考えよう。両方あれば，「A and B」という。どちらもない場合，「**not** A **or** B」という。一方，「not A and B」は，以下に示すように三つのケースが含まれる。

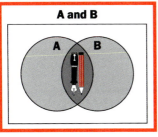

I always carry **a pen and a pencil** in my bag.

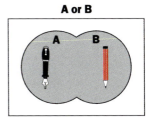

You can fill in the form with **a pen or a pencil**.

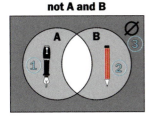

Today, I do **not** have **a pen and a pencil** in my bag because
① I have only a pen,
② I have only a pencil, or
③ I have nothing in my bag.

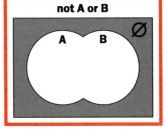

I cannot fill in the form because I do **not** have **a pen or a pencil**.

NOTE:「Without」も否定的な言葉なので，どちらもない場合に「without A or B」を使う。

　　　with a pen **and** a pencil　⇔　**without** a pen **or** a pencil

- This design does **not** depend on the size of the device (~~and~~) **or** the type of glass.
- One patient with brain damage could **not** draw objects from short-term memory (~~and~~) **or** recall the color of objects from long-term memory.
- We obtained a completely flat surface **without** any islands (~~and~~) **or** holes.
- Resist patterns rinsed with alcohol can be dried **without** collapse (~~and~~) **or** deformation.

142   Section 7

文章の主語の場合

○ **Both** houses **are** red.

○ A **and** B **are** red.

✕ *Both houses are not red.*
○ **Neither** house **is** red.

✕ *A and B are not red.*
○ **Neither** A **nor** B **is** red.

 ➡
Both ... are not → Neither ... is

Typical Mistakes

✕ *The front and rear facets of the laser were not coated.*
○ **Neither** the front **nor** the rear facet of the laser **was** coated.

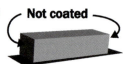
Not coated

✕ *Both facets of the laser were not coated.*
○ **Neither** facet of the laser **was** coated.

✕ *Both Solvent A and Solvent B cannot dissolve the base polymer.*
○ **Neither** Solvent A **nor** Solvent B **can** dissolve the base polymer.

✕ *When both pump pulses are not input into the comparator, the polarization of the probe pulse is preserved.*
○ When **neither** pump **pulse is** input into the comparator, the polarization of the probe pulse is preserved.

be consistent with

POINT:「**Consistent**」は「ほぼ等しい」、「ほぼ同じ」、または「よく一致している」を意味する言葉ではない。

DEFINITION: XはYと **consistent** であることは、XとYは矛盾していないという意味である。

○ The measurement results **are consistent with** the theory.
この文は結果が理論を支持するか、結果が正確であるか、または理論が正確であるということを意味するものではない。結果と理論がたがいに矛盾していないことだけを意味する。

- The measured energies dissipations (~~are almost consistent with~~) **are roughly equal to** the theoretical value of 1013 nJ.
- The values estimated by the two methods (~~are consistent with each other~~) **are almost the same**.

recover vs. restore

以前の良好な状態に戻る/戻す

Y recovers. (自動詞): 何かが **recover** すると，それは以前の良好な状態に戻る。「Recover」はこの意味で使用される場合，受動態はない。

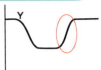

- The receiver cannot respond instantaneously to Signal 1. So, the signal initially deteriorates and then gradually **recovers**.
- When the stress time exceeds 300 hours, the collector current increases while the base current remains nearly constant. As a result, the current gain **recovers** at around 1,000 hours.

X restores Y. (他動詞): XがYを **restore** すると，XはYを以前の良好な状態に戻す。

- Multigigabit wireless links are useful for setting up a temporary network to **restore communications** after a disaster.
- Argon implantation causes deformation, and heating **restores the original shape** by making the argon desorb (脱着する).
- When the main **power supply is restored**, the rectifier begins to function.
- H_2 annealing **restores the hot-carrier reliability**.
- Half of **the loss in capacity** of a Li-ion cell **cannot be restored** by recharging.

何かを取り戻す

X recovers Y. (他動詞): XがYを **recover** すると，Xが失われた，または紛失しているYを取り戻す。

- The results show that the clock-and-data recovery circuit **recovered** the data and clock correctly.
- After a space shuttle launch, the booster rockets **are recovered** and used again.

PRACTICE: 空欄を「recover」か「restore」で埋めよ。

1. The device quickly _____ from the power-saving sleep mode in less than 10 μs.

2. Recharging a battery cell _____ the capacity lost by self-discharge.
3. Heating the liquid lowers the surface tension; but when it is cooled, the surface tension _____.
4. Adjusting the tuning current _____ the wavelength to the target value.
5. Due to slow reactions and weak muscles, it is often difficult for an elderly person driving an electric cart to _____ from a slide on slippery pavement.

monotonous vs. monotonic

POINT: ある日英辞書には「単調な」を「monotonous」とだけ訳されているが，技術英語で用いるのはこの単語ではなく，「monotonic」であることに注意しよう。

monotonous

DEFINITION: 何かが **monotonous** とは，それは常に一定なパターンを繰り返すことであり，非常に退屈というイメージが与えられる。

- Most factory jobs are very **monotonous**.
- The rain dripped **monotonously** from the trees.

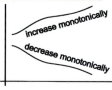
monotonic

DEFINITION: 何かが「**monotonically**+増加する」とは，どの値もその前の値より小さくないことを意味し，すなわち，どの点においても減少しないことである。また同様に，何かが「**monotonically**+減少する」とは，どの値もその前の値より大きくないことを意味し，すなわち，どの点においても増加しないことである。

- The standard deviation **increases monotonically**.
- …the **monotonic increase** in resistivity with annealing time…
- As the V/III ratio increases, the C^{12} concentration **decreases monotonically**.
- The insertion loss **rises monotonically** to −5 dB as the frequency increases from 40 MHz to 100 GHz.

Parentheses

左括弧の前にスペースを入れよう。ピリオド，コンマなどが後ろにある場合を除いて，右括弧の後ろにもスペースを入れよう。

1. リスト，ラベル，数値

- The composition, process state, and operating parameters strongly affect the quality indices (iron grade, basicity, tumbler index).
- There are three inputs (A, B, C) and two outputs (D, E).
- The sum of the threshold voltages of the n- and p-MOSFEETs (V_{tn} = 0.7 V, V_{tp} = –0.9 V) is greater than the supply voltage.

> リストは括弧で囲っている場合，「and」を使う必要はない。

2. 頭文字

✗ ...*Silicon-On-Insulator (SOI) technology*...
○ ...**s**ilicon-**o**n-**i**nsulator (SOI) technology...

✗ ...*the Equivalent-Input-Disturbance (EID) approach*...
○ ...the **e**quivalent-**i**nput-**d**isturbance (EID) approach...

> ただ頭文字が大文字で書かれているからといって，各単語の最初の文字も大文字にする必要はない。

3. 記号

- ...the threshold voltage (V_{th}) of the device...
- For each offset (k) with respect to the original sampling grid,...

> この場合，括弧の代わりにコンマを使ってもよい。
> - ...the threshold voltage, V_{th}, of the device...
> 一貫性を保つため，論文全体を同じスタイルで統一すること。

4. コメントと注釈

- A very large λ will satisfy all the timing constraints (if they can be satisfied) but will result in poor values of W.
- Each new generation (about every 3 years) of microprocessor has tended to be about three times faster than its predecessor.
- Compressed YIQ video uses a ping-pong scheme (one pair of memories for Y and one pair for IQ), providing...
- The results in Fig. 14 show the measured transfer function for no additional metal objects (squares), for the metal plate (triangles), and for the aluminum can (circles).

5. 参考文献，参照箇所関連

- In the 21064 processor (Fig. 3), the clock strip runs down the center.
- It corresponds to the method in which the hand model was calibrated (see [11]), which enables both the calibration and tracking procedures to...

Larger For A Than For B

> The X of A is larger than that of B.
> ⇨ The X is larger for A than for B.
>
> 最初の文のパターンは，多くの場合，第二文のパターンを用いて，より簡単に書くことができる。その結果，比較されているもの（AとB）がたがいにより近くに配置される。

- The modulation bandwidth <u>of a photonic system</u> is larger than <u>**that of an electronic one**</u>.
- ⇨ The modulation bandwidth is larger **for a photonic system** than **for an electronic one**.
- ⇨ The modulation bandwidth is larger **for a photonic** than **for an electronic system**.

- The transmission loss <u>of terahertz waves</u> in smoke is much smaller than <u>**that of infrared light**</u>.
- ⇨ The transmission loss in smoke is much smaller **for terahertz waves** than **for infrared light**.

- The lifetime <u>at $-270°C$</u> is about 100 times shorter than <u>**that at $25°C$**</u>.
- ⇨ The lifetime is about 100 times shorter **at $-270°C$** than **at $25°C$**.

PRACTICE: この文を「__er for A than for B」のスタイルに変換せよ。

1. The control performance for set A was better than that for set B.
   _____.

2. The etching rate of the In component is higher than that of the Ga component.
   _____.

3. The attenuation of the received power during snow was larger than that during rain.
   _____.

4. The peak intensity for the quantum wells is much larger than that for the GaN layer.
   _____.

5. The currents in an adiabatic SRAM are much smaller than those in a conventional SRAM.
   _____.

6. The multiplication factor at long wavelengths is larger than that at short wavelengths.
   _____.

7. The stability condition of Theorem 3 is more conservative than that of Lemma 2.
   _____.

8. The growth rate of the carbon-doped sample was smaller than that of the undoped sample.
   _____.

9. The allowable phase error for a 20-Gb/s demodulator is smaller than that for a 10-Gb/s one.
   _____.

10. The frequency characteristics of an optical modulator are flatter than those of a millimeter-wave mixer.
    _____.

11. The requirements for interchip interconnections are stricter than those for intrachip interconnections.
    _____.

Prepositions 7

空欄を適切な前置詞で埋めよ。必要がなければ，×で埋めよ。

REVIEW

a. X is incorporated _____ Y.
b. X arises _____ Y.
c. This paper is concerned ____ X.
d. As a result _____ X,...
e. X is made up ____ Y and Z.
f. This paper deals _____ X.
g. This paper discusses _____ a new approach.
h. X is different _____ Y.
i. X influences ____ Y.
j. X has an influence _____ Y.

CHECK YOUR KNOWLEDGE

1. X is inferior _____ Y.
2. X leads _____ Y.
3. X brings _____ Y. [= cause]
4. This paper focuses _____ X.
5. The upper/lower bound ____ Y...
6. _____ consequence,...
7. X is effective _____ doing something.
8. X is ____ accordance ____ Y.
9. oscillations ___ the conductance
10. I will talk _____ (a subject).

Section 8

fluctuations vs. variation

however, then, therefore, thus

with increasing frequency

almost

whose

performance vs. performances

flow

against

complete(ly) vs. perfect(ly)

summarize

reach

small AND red?

commercialized, specialized, standardized

compare between

express

XXXable

Change vs. Comparison

Punctuation: Slash, Capitals, Dash

Prepositions 8

fluctuations vs. variation

fluctuations

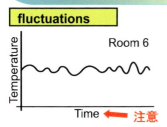

DEFINITION:「Fluctuations」は，一つのものについて使い，その特性の連続かつ不規則な変化を指す（規則的ならば，「oscillations」を使う）。「Fluctuations」は，ある特性が時間とともに変化していることを意味する。

- The **fluctuations in the temperature** of Room 6 are very small.
- The **temperature fluctuations** for Room 6 are very small.

NOTE:「Fluctuations」は通常複数形を用いる

- Datacom equipment should have a capacitor connected to the power line to suppress **voltage fluctuations**.
- The module has a temperature controller to reduce the change in wavelength due to **temperature fluctuations**.
- …**fluctuations in** the local magnetic field…
- The noise arises from **fluctuations in** the ground potential.

variation

DEFINITION:「Variation」は次のことを指す。
1. いくつかの事象に関してある特性の違い(Fig. A)。
2. 一方の変数の変化に従い，他方の変数が変化する(Fig. B)。
　いずれのケースにおいても，「variation」はよく最大値と最小値の差を指すのにも用いられる。

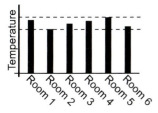

Figure A

- There is a small **variation in temperature** among the rooms.

- The **temperature variation** among the rooms is 7°C.

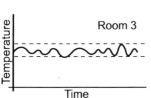

Figure B

- The **variations in temperature** for Room 3 are large.

- The **temperature variation** for Room 3 is large.

150   Section 8

- ○ The **variation in** received power among the channels is less than 1 dB.
- ○ After planarization, the **variation in** thickness is as small as ±0.43%.
- ○ The image is affected by noise and **local variations in** intensity and contrast.

TYPICAL MISTAKES

- ✗ *The limiting amplifier reduces packet-to-packet power fluctuations.*
  この文は一つのパケットのパワーではなくて，パケット間のパワーの違いに関するものなので，「fluctuations」は適切ではない。(パケット: 通信データの伝送単位)
- ○ The limiting amplifier reduces packet-to-packet power variations.

- ✗ *The roughness results in line-width fluctuations.*
  線の幅は時間とともに変化しない。
- ○ The roughness results in **variations in line width**.

- ✗ *The phase error is due to the fabrication fluctuation, for example, the width and depth of the mesa.*
- ○ The phase error is due to **process variations**, for example, variations in the width and depth of the mesas.

however, then, therefore, thus

POINT: 等位の語・句・節を対等の関係でつなげるときに，等位接続詞という単語を使う。このような単語は英語では七つある： and, but, for, nor, or, so, yet。「However」，「then」，「therefore」と「thus」は**副詞**であり，等位接続詞ではないので，二つの主節を接続するために使用してはいけない。

However: ✗   *Increasing the area makes the saturation current larger, however, it also increases the capacitance.*
              ○   Increasing the area makes the saturation current larger. **However,** it also increases the capacitance.

then: ✗   *In the absorption layer, the electric field becomes stronger, then the photoelectrons are accelerated.*
              ○   In the absorption layer, the electric field becomes stronger. **As a result,** the photoelectrons are accelerated.

therefore: ✗   *Increasing the data rate for mobile phones reduces the area covered by a base station, therefore more base stations are required.*
              ○   Increasing the data rate for mobile phones reduces the area covered by a base station. **Therefore,** more base stations are required.

thus:
- ✗ *The immunoassay takes around 4 min, thus the whole measurement procedure can be completed in under 10 min.*
- ○ The immunoassay takes around 4 min**.** **Thus,** the whole measurement procedure can be completed in under 10 min.

with increasing frequency

まず，以下の正しい英語を見てみよう。
○ The output power increases **with frequency**.

前置詞 — with ／ 前置詞の目的語 — frequency

その上で，次の文を考えてみよう。

○ The output power increases **with increasing frequency**.

同様に，この文において，「with」は前置詞であり，「frequency」は前置詞の目的語である。さて，「increasing」の品詞は何になるか？これは形容詞として使われている現在分詞である。以下の説明で次の文になぜ問題があるか理解できる。

✗ *The output power increases **with** increasi**ng** the frequency.*
　形容詞と名詞の間には「the」を入れてはならない。

✗ *The output power increases **with** increasi**ng** of frequency.*
　形容詞の後ろに「of」を入れてはならない。

NOT ENGLISH!

Typical Mistakes

- The thickness of the film decreases linearly **with increasing** (~~the~~) **pressure**.
- The degradation of the battery cell accelerated **with increasing** (~~of~~) **charging voltage**.

NOTE: 「With」のこの使用法を用いる場合，「with」の後ろの部分はいつも非常に簡単である。もっと複雑な表現をしたい場合は，代わりに「**as**」を用いた方がよい。

✗ *The bandgap of Si decreases with increasing strain induced by oxidation.*
○ The bandgap of Si decreases **as the strain** induced by oxidation **increases**.

✗ *The H_2 absorption improves with increasing degree of dispersion of the TiFeMn phase.*
○ The H_2 absorption improves **as the degree** of dispersion of the TiFeMn phase **increases**.

almost

「**Almost**」は，二つの状況を記述するために使用される。
1. ある事件が起こりそうで，起こらなかった。
2. 何かがある状態または状況に近づくが，まだその状態または状況にはなっていないこと。

要点は，「close to（接近している）」または「near（近い）」である。そのために，「almost」の同義語は「nearly」である。

Good Examples

O He **was almost hit** by a motorcycle.
オートバイにひかれそうになったが，**ひかれなかった。**

O The glass is **almost full**. ［完全にいっぱいになっていない。］

O It's **almost 5 o'clock**. ［まだ5時ではない。］

O The curve is **almost linear**.

O The power consumption is **almost zero**.

O These materials have **almost the same** density.

O The capacity remained **almost constant** at around 250 mAh/g.

Typical Mistakes

ほとんど+名詞　➡　almost all

- (~~Almost balls~~) **Almost all the balls** are red.
- The capacities of (~~almost the batteries~~) **almost all the batteries** decreased.
- (~~Almost Na~~) **Almost all of the Na** was replaced with Li.

almost mainly

- The lifetime of Li-ion cells for backup use is (~~almost~~) **mainly** determined by the degradation during storage, not by overcharging.

PRACTICE: 次の文中の誤りを見つけて直せ。

1. <u>Almost</u> DC networks contain a DC-DC converter.
2. The frequency is <u>almost</u> independent of bias voltage.
3. The contact current is <u>almost</u> 0 A.
4. The temperature of a coke oven is <u>almost</u> measured manually.
5. The operating speed is <u>almost</u> determined by the peak current density.
6. The spectrum in Fig. 5 <u>almost</u> satisfies the requirements.
7. The last stage of the amplifier <u>almost</u> outputs 1 V.
8. Automatic adjustment makes the output voltages <u>almost</u> constant.

whose

POINT: 技術英語では「**whose**」はあまり使わない。その代わりに，「with」か「which」を使ってみよう。

Pattern A: ~~whose X is ___~~ ➡ **with an X of ___**

*a voltage **whose** amplitude **is** more than 2 V*
⇨ *a voltage **with** an amplitude **of** more than 2 V*
*a coil **whose** self-inductance, L_0, **is** 1.2 μH*
⇨ *a coil **with** a self-inductance, L_0, **of** 1.2 μH*

PRACTICE A: 「Whose」を使わずに，次の表現を書き直せ。

1. a beam <u>whose</u> diameter is about 6 nm _____

2. a signal <u>whose</u> wavelength, λ, is 1,300 nm _____

3. pulses <u>whose</u> width is less than 30 ps _____

4. a layer <u>whose</u> thickness is 15 nm _____

5. an antenna <u>whose</u> gain, A_a, is 48.7 dBi _____

Pattern B: ~~whose X does Y~~ ➡ **with an X that does Y**

*a mask **whose** phase corresponds to the spherical aberration*
⇨ *a mask **with** a phase **that** corresponds to the spherical aberration*

PRACTICE B: 「Whose」を使わずに，次の表現を書き直せ。

1. signals <u>whose</u> frequencies extend down to several kilohertz
⇨ _____
2. a toothed antenna <u>whose</u> teeth correspond to frequencies from 150 GHz to 2.4 THz
⇨ _____
3. a converter <u>whose</u> capacitors are fabricated on the chip
⇨ _____
4. a package <u>whose</u> coefficient of thermal expansion is close to that of $LiNbO_3$
⇨ _____
5. an antenna <u>whose</u> length is half the wavelength of the carrier signal
⇨ _____
6. foam cubes <u>whose</u> edges are 25 cm long
⇨ _____

154   Section 8

Pattern C: ~~whose~~ ➡ which

...this material, <u>whose</u> melting point is 32°C,...
⇨ ...this material, **which** has a melting point of 32°C,...

...a JN-60 coke oven, whose combustion chambers are 6 m high,...
⇨ ...a JN-60 coke oven, **which** has combustion chambers **that** are 6 m high,...

PRACTICE C: 「Whose」を使わずに，次の表現を書き直せ。

1. fixed wireless access, <u>whose</u> top speed is 622 Mb/s,

⇨ _____

2. The image rejection ratio is 49 dB, <u>whose</u> value satisfies the specifications for short-range wireless systems.

⇨ _____

3. the module, <u>whose</u> size is the same as that of a standard LD module,

⇨ _____

4. We used Kovar, <u>whose</u> characteristics are close to those of the glass.

⇨ _____

performance vs. performances

単数形： 技術英語では，複数のパフォーマンス・インデックスが言及されても，「performance」は大抵**単数**である。（性能）

複数形： 複数形の「performance**s**」は，一般的に音楽家，演劇グループまたは芸能人が話題になっているとき用いられる。（興行，演奏）

Typical Mistakes

- The (~~performances~~) performance of the wireless link (~~are~~) is now being tested.

- The final topic is the bit-error-rate (~~performances~~) performance.

- (~~These performances are~~) This performance is good enough to meet IEEE specifications.

- The heat cycle (~~performances were~~) performance was examined at operating temperatures from 20°C to 100°C.

flow

POINT: 図は flow ではなく，手順，プロセスまたはフローチャートを示すものである。すなわち，日本語の「フロー」を通常「flow」と直訳してはならない。

~~flow~~ ➡ **procedure**

BASIC PATTERNS
In the XXX procedure (Fig. 3), first …
Figure 3 illustrates/shows the XXX procedure.
Figure 3 show the steps in the XXX procedure.

Step 1
Step 2
Step 3
Step 4

✗ *Figure 3 shows <u>the flow of image adjustment</u>.*
○ Figure 3 illustrates **the image adjustment procedure**.

✗ *Figure 6 is a diagram of <u>the conventional testing flow</u>.*
○ Figure 6 shows **the steps in the conventional testing procedure.**

✗ *Figure 8 shows <u>a flow of matching steps</u> performed in the processing array.*
○ Figure 8 illustrates **the matching procedure** performed by the processing array.

✗ *<u>The flow of measurements</u> is shown in Fig. 5. First, electrons are stored in the memory node.*
○ In **the measurement procedure** (Fig. 5), first, electrons are stored in the memory node.

~~flow~~ ➡ **flow chart**

BASIC PATTERN
Figure 3 shows a flow chart of _____.

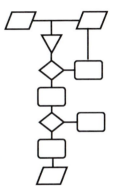

✗ *Figure 7 shows a <u>flow</u> of the image enhancement algorithm.*
○ Figure 7 shows a **flow chart** of the image enhancement algorithm.

✗ *Figure 6 is a diagram of the conventional testing <u>flow</u>.*
○ Figure 6 shows a **flow chart** of the conventional testing procedure.

156   Section 8

~~flow~~ ➡ **fabrication process**

BASIC PATTERNS
Figure 6 **illustrates/shows the fabrication process**.
Figure 6 **illustrates the procedure for fabricating** XXX.
Figure 6 **shows the (main) steps in the fabrication process**.
Figure 6 **shows the (main) steps in the fabrication of** XXX.

- ✗ *Figure 4 shows the flow of a fabrication process.*
- ○ Figure 4 shows **the fabrication process**.
- ○ Figure 4 shows **the steps in the fabrication process**.

- ✗ *Figure 5 shows the sensor fabrication process flow.*
- ○ Figure 5 illustrates **the procedure for fabricating a sensor**.

- ✗ *Figure 9 shows the outline of process flow to fabricate buried optical waveguides.*
- ○ Figure 9 shows **the main steps in the fabrication of** a buried optical waveguide.

against

~~Y against X~~

POINT:　一般的に，二つの変数間の関係を表す場合には，「Y against X」という表現は使えない。このとき，次の表現が使える。

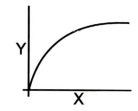

Figure 3 shows...
We measured...
{
　Y **versus** X.
　how Y **varies with** X.
　how Y **changes with** X.
　how Y **depends on** X.
　Y **as a function of** X.
　the dependence of Y **on** X.
　⋮
}

順番に注意：
Y軸変数が，最初に書かれる。

Figure 3 is a graph of Y **versus** X.

例外: **plot**（動詞）
- In this graph, Y **is plotted against** X.
- We **plotted** Y **against** X.

complete(ly) vs. perfect(ly)

Basic Concept

あまりよい状態にないが，このカップと受け皿のセットは **complete**（完全）である。すべての部分は存在するし，不足しているまたは**なくなっているものはない**。

一部だけが存在するので，このセットは **partial** または **incomplete** である。すなわち，**何かが欠けている**または**不足している**。

このセットは **perfect** である。**完全**だけでなく，エラー，損害，**欠陥などもない**。上の両方のセットは **imperfect** である。

「**Complete(ly)**」は，何が可能な範囲に極限まで起こるということを意味することもある。「**Partial(ly)**」は，制限に達していないということを意味する。

- This structure **completely/partially eliminates** electrical interference in a module.
- ...the **complete/partial elimination** of electrical interference...

- Heat treatment **completely/partially removes** the hydrogen.
- ...the **complete/partial removal** of the hydrogen...

- The film **completely/partially covers** the electrodes.
- ...the **complete/partial coverage** of the electrodes ...

- The load capacitance **completely/partially suppresses** voltage oscillations.
- ...the **complete/partial suppression** of voltage oscillations...

「**Perfect(ly)**」は，異常なふるまいや，エラーがないという意味に使われることもある。

- Polyimide **works perfectly** as a **protective** layer.
- ...the **perfect protection** provided by the polyimide layer...
- The PMMA domain **is perfectly aligned** with the edge of the guide pattern.
- ...the **perfect alignment** of the PMMA domain with the edge of the guide pattern...

Section 8

次の文で,「**perfectly**」と「**completely**」を比較しよう。

To the greatest extent; fully — **No error**

- The disturbance **was almost perfectly estimated**, and the improved control law **almost completely compensated for** the dead zone. The output **tracked** the reference input **almost perfectly** (steady-state error: 0.03).

PRACTICE: 空欄を「complete(ly)」か「perfect(ly)」で埋めよ。

1. A repetitive control system can _____ track a periodic reference input.
2. The material _____ fills the holes.
3. The center of the wire is _____ oxidized.
4. The temperature dependence shows _____ agreement with the theoretical prediction.
5. Noise is _____ filtered out.
6. These properties are a _____ match to the characteristics required for integrated photonics.
7. There are two problems that have not been _____ solved.
8. The optical paths are _____ inside the prisms.

Completely + Adjective (形容詞)

何かが可能な範囲内にほぼ限界に達している特性を強調するために,「completely」は形容詞とともに使用できる。「Perfectly」はいくつかの形容詞とともに使用できるが,その数は限られている。

completely flat	perfectly flat
completely dry	perfectly dry
completely wet	~~perfectly wet~~
completely round	perfectly round
completely different	~~perfectly different~~
completely compatible	perfectly compatible
completely negative	~~perfectly negative~~
completely independent of	~~perfectly independent of~~

強調のために

completely different   **exactly** the same   **(absolutely) identical**

- The waveguides have **completely different** structures.
- The transmission characteristics of terahertz waves in smoke are **completely different** from those of infrared light.
- Our new device is **exactly the same** size as a conventional one.
- The photodiode and the fiber output exhibit **exactly the same** attenuation characteristics.

- The spectra for the two ports are **identical**.
  「Identical」は「exactly the same」を意味するので，通常それを強調する必要はない。しかし，二つのものは異なると予想しているのに，実際にそれらが同じであるとわかって驚いた場合，「absolutely」という単語を加えることによって非常に強い強調を示すことができる。
- The spectra for the two ports are **absolutely identical**.

summarize

POINT: 表（テーブル）を説明するために，「**summarize**」を使わないこと。「**List**」という動詞を使うべき。

　何かをsummarizeするときには，新情報を提示しない。ただ前に述べた話の重点を短く繰り返す。一般に，表はものを **list** するし，図はものを **show** する。

Typical Mistakes

- Table 5 (~~summarizes~~) **lists** the measurement results.
- Table 4 (~~summarizes~~) **lists** the specifications of the radar module.
- The simulation conditions (~~are summarized~~) **are listed** in Table 4.

✗ *The ratios of f_r at 85°C to that at 25°C <u>are summarized</u> in the table.*
○ The table **lists** values of f_r at 85°C normalized by the value at 25°C.

✗ *In Fig. 10, the graph <u>summarizes</u> the N_s and mobility of InP-based MD structures in this work and in the literature.*
○ The graph of mobility versus N_s for InP-based MD structures (Fig. 10) **shows** data from this study and from the literature.

Good Examples

- The above **discussion can be summarized** in a design algorithm for the modified repetitive-control system.
- [1] **summarizes the work** conducted up to around 1960.
- [*Oral presentation*]　Finally, I'll summarize my presentation.

reach

POINT:「**Reach**」は他動詞である。その後ろの単語は，前置詞の「to」ではなくて，目的語である。

- The start-up converter starts a charge-pump operation when V_{in} **reaches** (~~to~~) **0.5 V**.
- The system **reached** (~~to~~) **the steady state** in the 6th period.
- Combustion progresses until it **reaches** (~~to~~) **the bottom** of the bed.
- Investment in renewable energy **reached** (~~to~~) **a new record** in 2011.

small AND red?

small houses and large houses = small and large houses

small houses and red houses = small and red houses

small houses that are red = small red houses

POINT 1: 名詞の前にある複数の形容詞は通常「and」でつながないこと。

- We devised a <u>simple scalable</u> architecture.
- We need <u>faster more compact</u> systems.
- <u>Continuous uniform</u> flow is needed for an accurate immunoassay.
- …<u>high-speed high-accuracy</u> channel switching…
- <u>Smaller cheaper optical</u> transceivers are necessary.
- This system ensures <u>continuous safe</u> operation of the sintering process.
- …a <u>simple cost-effective</u> base station...

例外： 決まった表現であれば，名詞の前でも複数の形容詞を「and」でつないでもよい。

- ...a <u>tried and true</u> method... (= 頼りになる)
- There is no <u>hard and fast</u> rule. (= 非常に厳重な，がんじがらめの)
- ...a <u>long and winding</u> road...

POINT 2: 同じ種類（色，大きさ，重さなど）の形容詞は名詞の前に「and」でつないでもよい。

- ...a <u>red and yellow</u> house...
- ...<u>small and large</u> houses...
- This amplifier has <u>positive and negative</u> outputs.
- It exhibits good <u>static and dynamic</u> performance.
- A comparison of the <u>measured and calculated</u> results reveals that…
- The <u>red and blue</u> lines show the collector and base currents, respectively.
- This laser is a promising candidate for <u>medium and long</u> distances.

DANGER: 「High-speed and low-power LSI」は不自然な英語なのに，「LSI」は単数なので，解釈は一つしかなく，二つの特徴を持っている一種のLSIである。しかし，「LSI」に「s」をつけて複数形にすると，「high-speed and low-power LSIs」になる。最も自然な解釈で，この表現は二種類のLSIを意味することになるので，注意すべきである。

commercialized, specialized, standardized

POINT: 通常これらの単語は形容詞ではなく，**動詞**として使用される。

commercialize

動詞 | **DEFINITION:** 最近開発された製品やサービスを **commercialize** する（商品化する）ならば，それを**公開し市場で売り始める**という意味である。

○ **To commercialize** 10G-EPON systems in the near future, we need a small, inexpensive optical transceiver.
○ 100G technologies **have already been commercialized**.

形容詞 部品または装置を店で買うことができる，または，それを自社で開発したことよりも，買ったことを明らかにしたい場合，「**commercialize**」を使用してはいけない。

~~commercialized~~ ➡ **commercial**
commercially available

- Figure 8 shows data for a (~~commercialized~~) **commercial** 100G modulator.
- A (~~commercialized~~) **commercial** thermoelectric cooler can easily handle temperature variations of this magnitude.
- (~~Commercialized~~) **Commercially available** sensors have many good characteristics.
- A comparison of two electric-field sensors, a (~~commercialized~~) **commercially available** one and our new optical one, revealed that…

specialize

動詞 技術英語では，通常「specialize」という動詞はあまり使用しない。

形容詞 ~~specialized~~ ➡ **special**

- A (~~specialized~~) **special** learning procedure is needed to tune the classifiers.
- The misalignment of the cores is smaller for this (~~specialized~~) **special** fiber than for standard single-mode fiber.
- The tumbler index is determined by first placing a sample of sintering agglomerate in a (~~specialized~~) **special** tumbler and rotating it for 8 min.
- This technique enables the design of such systems without a (~~specialized~~) **special** design tool or a high-performance calculator.

standardize

動詞 | **DEFINITION:** 「**Standardize**」の一つの意味は**標準（規格）を決める**ことである。

Section 8

○ The protocol for a 10-gigabit Ethernet passive optical network (10G-EPON) was standardized in 2009 as IEEE802.3av.

形容詞　~~standardized~~　➡　**standard**

- The (~~standardized~~) **standard** specification contains two transmission ranges.
- A (~~standardized-size~~) **standard-size** F-band waveguide was connected to Port 1.

✗ compare between

POINT: 「To **compare between** X and Y」は正しい英語の表現ではない。しかし、「to **make a comparison between** X and Y」はOKである。

~~compare between Y and Z~~
~~compare X between Y and Z~~　➡

compare Y and/with Z
compare the X of Y and/with that of Z

Typical Mistakes

✗ *We compared the roughness between Resist A and Resist B.*
○ We **compared** the roughness of Resist A **and** that of Resist B.

✗ *This function compares the bits between the optical label and the local address.*
○ This function **compares** the bits of the optical label **with** those of the local address.

✗ *The characteristics were compared between two types of diodes.*
○ The characteristics **of** two types of diodes **were compared**.

✗ *Figure 7 compares the etching rate between two conditions.*
○ Figure 7 **compares** the etching rates **for** two conditions.

Good Examples
comparison

○ Changing β strongly affects control, as can be seen from **a comparison between/of** (b) **and** (c) in Fig. 6.
○ Table 2 shows **a comparison between/of** a Li ion capacitor **and** a Li ion battery.
○ **A comparison of** our new method **with** conventional methods revealed the following points.
○ **A comparison of** $C_{g(Tx)}$ **and** $C_{g(Rx)}$ allows us to determine the validity of the estimates.

express

> **POINT 1:** 通常，何かのための式や関係を導入するために，「**express**」という単語を使用してはいけない。「**Be given by**」または他の表現を使った方がよい。

be expressed by ➡ **be given by, be-動詞**

- The relaxation oscillation frequency (~~can be expressed by~~) **is given by**
$$f_R = \frac{1}{2\pi}\left[\frac{1}{\tau\tau_p}\left(\frac{I}{I_{th}}-1\right)\right]^{1/2}.$$

- The impedance, Z, (~~is expressed as follows:~~) **is given by**
$$Z = \frac{1}{j\omega C_b + j\omega L_{pr}} \ . \tag{4}$$

- The circuit equations of the equivalent circuit (~~are expressed as follows:~~) **are**
$$V_b = V_g + V_s + (R_s + jX_v)\{-j\omega C_b V_b - j\omega C_{sb}(V_b - V_g)\} \tag{9}$$
and
$$j\omega C_g V_g + j\omega C_b V_b = 0 \ . \tag{10}$$

> **POINT 2:** モデルまたはシミュレータにおいて，特徴を describe または represent する。この場合，「express」という単語はあまり使わない。

~~express~~ ➡ **describe / represent**

- All these blocks of the optical receiver can be (~~expressed~~) **described/represented** in a conventional electrical-circuit simulator.
- A technique for (~~expressing~~) **describing** wavelength in an electrical-circuit simulator needs to be developed.
- This model cannot (~~express~~) **describe** the different impacts that the variables L and E have on Y.

> **POINT 3:** 「Express」はあるものが何か**特定の数学的な形式**で書けるためにしばしば用いられる。

Good Examples

○ Goldbach's conjecture, an unsolved problem in number theory, states that every even integer greater than 2 **can be expressed as** <u>the sum of two primes</u>.

○ A fraction **can be expressed as** <u>a repeating decimal</u>.

Section 8

○ Relative humidity **is usually expressed as** a percentage:
$$\phi = \frac{e_w}{e_w^*} \times 100\,\%\,,$$
where e_w is the partial pressure of water vapor and e_w^* is the saturated vapor pressure of water.

○ Electrical resistance is defined to be the ratio of voltage to current: $R = V/I$. It **can also be expressed in terms of** the resistivity, ρ: $R = \rho\,(l/A)$, where l is the length of the material and A is the cross-sectional area.

○ In the metric system, parts per million (ppm) **can be expressed in terms of** milligrams and kilograms: 1 ppm = 1 mg/kg. It **can also be expressed as** a percentage: 1 ppm = 0.0001%.

XXXable

以下の表現のどちらが正しいか？

   tunable-wavelength laser   tunable-temperature laser
   wavelength-tunable laser   temperature-tunable laser

答えは以下の通りである。

基本的な意味

通常，動詞と「-able」接尾辞を結合することによって作られる形容詞では，動詞は**受動的な意味**を持っている。次の例を吟味しよう。

usable	⇨	able to **be used**
measurable	⇨	able to **be measured**
movable	⇨	able to **be moved**
tunable	⇨	able to **be tuned**
polarizable	⇨	able to **be polarized**

Pattern 1:「XXXable―名詞」

単純な「形容詞―名詞」のパターンから始めよう。

 The voltage is high. ⇨ high voltage
 a generator outputting a <u>high voltage</u> ⇨ a **high-voltage** generator

「XXXable―名詞」のパターンは同じであり，「XXXable」という単語が，**単純な形容詞**として使われる。

 The speed is variable. ⇨ variable speed
 a drive with a <u>variable speed</u> ⇨ a **variable-speed** drive
 The wavelength is tunable. ⇨ tunable wavelength
 a laser with a <u>tunable wavelength</u> ⇨ a **tunable-wavelength** laser

他の例：

  programmable-logic controller   **variable-gain** amplifier
  renewable-energy systems   **adjustable-length** trekking poles

Pattern 2: 名詞—XXXable

簡単な文章を先に作り、それを変換してみよう。
A machine (= computer) is able to read the data.
The data is able to be read by a machine.
The data is **readable** by a machine.

さて、Section 4の「**Adjective Formation (-ed)**」の題の例を思い出してみよう。

The system is powered by a battery.
⇨ a **battery-powered** system （電池給電式のシステム）

上記の文を、まったく同じように変換できる。

The data is readable by a machine.
⇨ **machine-readable** data （機械可読データ）

このパターンでは、「XXXable」という単語は、単純な形容詞として使わず、**受動態動詞の特性**も持っている。

他の例: The laser is tunable by means of temperature.
⇨ a **temperature-tunable** laser
The adhesive is curable by UV (light).
⇨ **UV-curable** adhesive （UV硬化性接着剤）
The steel is treatable by heat.
⇨ **heat-treatable** steel （熱処理可能な鋼）

分析

Ex. 1: 「*Friction-controllable fluid bearing*」と言う表現は正しいであろうか？この表現は今までの説明によると、以下のことを意味する。
• a fluid bearing that is controllable by friction

それは意味を持たない。流体軸受は、摩擦によって制御されるのではなく、摩擦を制御する。Section 4の「**Adjective Formation (-ing)**」の題の例によって、「*the diode emits light*」という文から「*light-emitting diode*」が出る。それと同じように、「*the fluid bearing controls the friction*」から以下の表現が出る。
• **friction-controlling** fluid bearing

Ex. 2: 「Refractive-index-controllable SiO_x」という表現は正しいだろうか？上記の議論に基づいて、それは以下を意味する。
• SiO_x that is controllable by the refractive index

この表現も不適切である。屈折率は制御可能である。すなわち、「*the refractive index is controllable*」。「*Controllable*」は単純な形容詞として使われているので、「*controllable refractive index*」は正しい。したがって、上記のPattern 1を使用すべきである。
• **controllable-refractive-index** SiO_x

Change vs. Comparison

POINT: 「Increase」、「decrease」、「reduce」、「improve」および「enhance」という言葉は常に**変化する**ことを表す。二つの異なるものを比較するときに、それらを使用してはいけない。

Ex. 1: 古い黄色の鉛筆があり、新しい赤い鉛筆を買う場合、次の文は正しいか？
- The new pencil has <u>increased</u> length.

答えはnoである。「Increase」という言葉は常に変化することを表すが、2本の鉛筆のいずれにも長さに**変化がない**。二つの異なるものを比較するために、形容詞の比較形を使用すること。
- The new pencil is longer than the old one.
- The new pencil has greater length than the old one.

一方、鉛筆削りを使い、赤い鉛筆を削った場合、次の文は正しいか？
- The red pencil has <u>reduced</u> length.

削ることによって、鉛筆の長さが**変化した**ので、答えはyesである。この文は、2本の異なる鉛筆を比較していないことに注意する。

Ex. 2: エンジンAは、100馬力を出力する。150馬力を出力する新しいエンジン（エンジンB）が造られたとすると、次の文は正しいか？
- Engine A has <u>reduced</u> power.
- Engine B has <u>increased</u> power.

答えはnoである。この2台のエンジンが製造された後、その出力は**変化しない**ので、次の文は正しい。
- Engine B outputs more power than Engine A.
- Engine B is more powerful than Engine A.

なお、エンジンAを搭載した車の最高速度は、150 km/hであった。その後、エンジンAがエンジンBに交換され、最高速度は、200 km/hとなった。次の文は正しいか？
- The car has <u>increased</u> speed.

車は同じであり、最高速度が**変更した**ため、答えはyesである。

Typical Mistakes

✗ *The new laser structure provides <u>increased</u> output power compared to a conventional one.*

○ The new laser structure provides **greater** output power than a conventional one.

- The use of a triple mesa results in a (~~reduced~~) smaller dark current on the surface.

Slash

　スラッシュの使用に関する問題は，意味が厳密に定義されていないことである。通常，スラッシュは「or」を意味する。しかし，例えば「washer/dryer」や「clock/radio」のように，二つのものが結合してできたものであるか，または二つの機能を持つ場合，スラッシュは「and」を意味する。スラッシュの意味は厳密に定められていないので，**標準的な専門用語以外には，使うのを避けた方がよい**。例えば，次の例を考えよう。

✗ *The circuits might undergo physical/chemical degradation.*
この文章には色々な解釈がある．
   A. The circuits might undergo either **physical or chemical degradation, but not both**.
   B. The circuits might undergo **physical or chemical degradation, or both**.
   C. The circuits might undergo **both physical and chemical degradation**.

これほどの曖昧さは技術論文にふさわしくない。「or」という意味なら，「or」という言葉自体を書いた方がよいし，「and」という意味なら，「and」という言葉自体を書いた方がよい。そして，両方を意味するつもりならば，「and/or」を書くべきである。

○ The circuits might undergo physical **and/or** chemical degradation.
○ The circuits might undergo physical **or** chemical degradation**, or both**.

Typical Mistakes

- …multilayers formed by (~~ion-beam/helicon~~) **ion-beam or helicon** sputtering.
- …receiver configurations for (~~analog/digital~~) **analog and digital** optical transmission systems.
- …good accessibility to (~~data/program~~) **data and program** resources…

専門用語では，スラッシュは色々な意味がある。
通常，スラッシュの前後にスペースを入れないことに気をつけよう。

And	**I/O** circuits　　(input **and** output) fingerprint **sensor/identifier** LSI　　(sensor **and** identifier) the **read/write** functions of a RAM　　(read **and** write)
To	**A/D** converter　　(analog-**to**-digital) **O/E** conversion　　(optical-**to**-electrical)
Structure	**Si/SiO$_2$** interface　　(the interface between Si and SiO$_2$) **CMOS/SOI** circuit　　(CMOS circuit on an SOI wafer)
Ratio	**ON/OFF** ratio　　(ratio of voltage of ON signal to that of OFF signal) **S/N** ratio　　(signal-to-noise ratio) (Also, SNR.)

With & without	チャートやグラフにラベルを書くスペースが少ない場合，「**with**」を「**w/**」，また，「**without**」を「**w/o**」と省略して書くことができる。 　　　　w/ bias　　　w/o bias

Capitals

1. 「Section」と「chapter」の後ろに数値がある場合，これらの単語の頭文字を大文字にすること。

○ As mentioned in Section 2, the main problem …
○ …(see Chapter 5).

2. 節の見出しの頭文字を大文字にする。

○ 1. Introduction
○ References

3. 化学物質および元素の名前をフルネームで書く場合，頭文字は大文字にしないこと。

○ Annealing was carried out in a nitrogen (N_2) ambient.
○ The n-type dopant was silane (SiH_4) and the p-type was diethylzinc (DEZn).

4. ただ略語および記号が大文字で書かれているからといって，そのものの頭文字も大文字にする必要はない。

✗ *Solid-Oxide Fuel Cells (SOFCs) are attracting a great deal of attention ...*
○ Solid-oxide fuel cells (SOFCs) are attracting a great deal of attention ...

✗ *...a new Voltage-Controlled Oscillator (VCO)...*
○ …a new voltage-controlled oscillator (VCO)…

Dash

NOTE: ダッシュは技術英語であまり使われていない。その機能はコンマと括弧により実現できる。通常ダッシュの前後にはスペースを入れないことに気をつけよう。

1. リスト

○ The complete circuit—quadrature frequency divider, dual interpolators at $f/2$, and an XOR frequency doubler—performs the function of a phase shifter running at the full frequency of the system clock.

2. コメント

○ One or both of the above effects can be expected in many—if not most—relevant experimental situations.
○ Behavioral VHDL models of burst-mode state machines can be very lengthy—over 1,600 lines of VHDL code.

Prepositions 8

空欄を適切な前置詞で埋めよ。必要がなければ，×で埋めよ。

REVIEW

a. X is the same _____ Y.

b. X is effective _____ improving Y.

c. There is no information _____ Y.

d. X has an effect _____ Y.

e. X brings _____ Y. [= cause]

f. X is _____ the order _____ 10^6.

g. X leads _____ Y.

h. X compensates _____ Y.

i. X is called _____ Y.

j. X takes Y _____ consideration.

CHECK YOUR KNOWLEDGE

1. In Fig. 1, the diagram _____ the top is …

2. In Fig. 2, the diagram _____ the bottom is …

3. In Fig. 3, the diagram _____ the left is …

4. In Fig. 4, the diagram _____ the right is …

5. In Fig. 5, the diagram _____ the middle/center is …

6. The key _____ X is…

7. There is a difference _____ X and Y.

8. There is a difference _____ length _____ X and Y.

9. The best approach _____ X…

Section 9

口頭発表のヒント

形式のレベル

初めに

アウトライン

次のトピックへの移行

聴衆を見ること

I vs. We

略語の導入

スライド

レーザーポインター

終わり

暗　記

プレゼンテーションを短くする方法

質疑応答の時間

練　習

セッションの司会を務める方法

初めに　171

形式のレベル

プレゼンテーションの英語は少しフォーマルでなくても構わない。口語でもOKであるし，ときにはジョークでも構わない。

初めに

Name & Affiliation (名前と所属)

通常，名前と所属を言う必要はない。
× *"My name is Taro Suzuki. I work at the ABC Lab of the XYZ Company."*.

まず，聴衆の誰もが見ている会議プログラムに名前が書かれている。第二に，司会者は発表者を紹介するときに名前をいう。第三に，名前はプレゼンテーションのタイトルのスライドに記載されている。この三つで十分である。

Thank you (お礼)

お礼でプレゼンテーションを開始すること。

1. 普通のプレゼンテーション（招待されたプレゼンテーションではない）

短く，シンプルにした方がよい。

- "Thank you for the introduction. I'll be talking to you about [*subject of your paper, not the title*]."
- "Thank you, Professor Johnson. Good morning, everyone. My talk concerns [*subject of your paper, not the title*]."

NOTE:「Thank you, Mr. Chairman.」と「Thank you, Chairperson.」という文章は，技術的な会議にはあまりにも形式的であり，通常は英語のネイティブスピーカーは使用しない。

2. 招待されたプレゼンテーション

セッションの司会者に感謝した後，組織委員会，プログラム委員会，または運営委員会に招待へのためにお礼をいうこと。

- "Thank you for the introduction. It's an honor and a pleasure to be here, and I'd like to thank the organizing committee for giving me this chance to discuss our recent work. I'll be talking to you about [*the fabrication of MEMS mirrors…*]."
- "Thank you, Dr. Spencer. First of all, it's an honor to be here; and I'd like to thank the program committee for inviting me. In this talk, I'll discuss [*the potential of photonics technologies…*]."
- "Thank you for the introduction. It is a great honor and a pleasure to be here, and I would like to thank the program committee for inviting me to this prestigious conference. I'll be talking about [*an optical transmitter for…*]."

アウトライン

　タイトルのスライドの後，ほとんどの発表者は，プレゼンテーションのアウトラインを示す。これは，可能な限り単純にすべきである。目的は，研究の詳細を説明することではなく，単にトピックの順序を示すことである。

　また，アウトラインの説明もシンプルにするべきである。例えば，右側のアウトラインについては，次のことを言ってもよい。

> **Outline**
> Background
> Device structure
> Fabrication process
> Experimental setup
> Measurement results
> Summary   2

- "After giving you a little background, I'll explain the device structure and the fabrication process. Then, I'll describe the experimental setup and present some measurement results. (And I'll finish up with a summary.) Now, let's begin with the background."

NOTE: 誰もがプレゼンテーションの最後にまとめを述べるので，アウトラインを説明するときには，「Summary」を述べる必要はない。しかし，それを述べたいなら，述べてもよい。

次のトピックへの移行

　プレゼンテーションの際に，発表者の説明がアウトライン中の一つのトピックから次のトピックに移るときには，聴衆に断わるべきである。

- "Next, I'll explain the experimental setup."
- "The next topic is the experimental setup."
- "Now, let's look at the experimental setup."
- "Before getting into the results, I'll describe the experimental setup."

　これらの断わりをいうとき，再びアウトラインを示した方がよい。次の項目は赤で強調表示するか，他の項目はグレーで表示すべきである。

> **Outline**
> Background
> Device structure
> Fabrication process
> **Experimental setup**
> Measurement results
> Summary   9

> **Outline**
> Background
> Device structure
> Fabrication process
> **Experimental setup**
> Measurement results
> Summary   9

聴衆を見ること

　話すときに，聴衆を見ることによって，よりよい印象が与えられる。全体のプレゼンテーションを暗記していない場合でも，最初の三，四枚のスライドの説明内容を暗記した方がよい。そうすると，話しながら聴衆を見ることが容易になる。何かを指し示すときだけに，スライドを見るべきである。**スライドの目的は，発表者がいうことを思い出させるためのものではない**。それは聴衆が理解する手助けをするものである。

I vs. We

　プレゼンテーションで「I」と「we」をともに使用することができる。しかし，使用方法について注意する必要がある。

研究について

　研究は，複数の人によって行われた場合は，その研究を説明するとき，必ず「we」を使用すること。

- ✗ *"I fabricated these devices by…"*
- ○ **"We fabricated** these devices by…"
- ✗ *"I analyzed…"*
- ○ **"We analyzed**…"

プレゼンテーションについて

　自分一人でステージ上で話しているので，プレゼンテーション自体について話すとき，「I」を使用すること。

- ✗ *Next, we will explain the experimental setup.*
- ○ "Next, **I'll explain** the experimental setup."

略語の導入

　英語の文章では，略語を導入するときは，括弧にいれる。
- The power is adjusted with a variable optical attenuator (VOA) using analog feedback.

　口頭発表では，「or」という単語を使用して略語を導入する。
- "We adjust the power with a variable optical attenuator, or V-O-A, using analog feedback."

　「or」＋略語の前後に少し間をおいた方がわかりやすい。

"…a variable optical attenuator… or V-O-A, …using analog feedback"

NOTE: プレゼンテーションで一度か二度しか使用しない場合は，略語を導入する必要はない。その表現自体を使用すべきである。

スライド

A. 簡単にせよ！（Keep It Simple!）

1. 説明しないことを見せないこと！ プレゼンテーションで話さない情報を，スライドに表示してはいけない。スライドは，可能な限り単純にすべきである。スライドに多くのテキストや他の情報を詰め込むべきでない理由は以下である。

- ◇ スライドの内容について，聴衆はテキストより絵を好む。
- ◇ 聴衆は大量のテキストを読む時間がない。
- ◇ スライドに情報が多すぎると理解しにくくなる。スライドは理解しやすくする必要がある。

2. スライドは観衆のためのものであって，発表者のためのものではない。 スライドは，発表者の記憶補助ではない。いうべきことを思い出すための文をスライド上にそのまま書いてはいけない。聴衆は，発表者がスクリーンに表示されている文章をそのまま棒読みすることを好まない。

B. 文字の大きさ

会議室の後ろの席で読めるように，すべての**文字が十分に大きい**ことを確認すること。

C. タイトル

スライドのタイトルは，単にスライドの内容を記述するよりはむしろ，スライドの**要点**またはスライドの**意味**を伝えた方がよい。

D. 図中の線のラベル

できれば，線のラベルを，線自体のすぐそばに書いた方が，聴衆に理解しやすい。色も役に立つ。

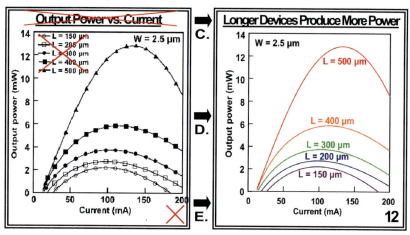

E. スライドの番号

スライドの番号を右下か右上の角に置く。質疑応答の時間に，その番号は質問者がスライドを特定するのに役に立つ。

F. 図と数式の番号

図または数式の番号をいう必要はない。
- ✗ *"Figure 3 shows…"*
- ✗ *"…as shown in Equation 4."*

図または数式を指して，次のような表現をいう。
- ○ "This graph shows…"
- ○ "…as you can see here."

あるいは，指すことの有無にかかわらず，以下のような言葉をいうことができる。
- ○ "The graph on the left shows…"
- ○ "The equation at the bottom indicates that…"

G. 行頭記号

行頭記号としてハイフンを使ってはいけない。

まず，行頭記号として使われるハイフンは美しくない。次に，項目が数字から始まるならば，ハイフンはマイナス記号と間違える。それは聴衆の理解を妨げる。

情報を明らかに示すのに，時にはインデントだけで十分である。もし行頭記号が必要であると思えば，小さな黒い点または他のシンボルを使った方がよい。

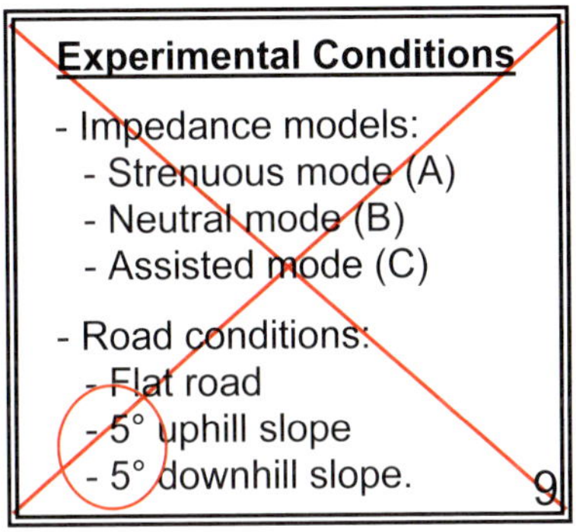

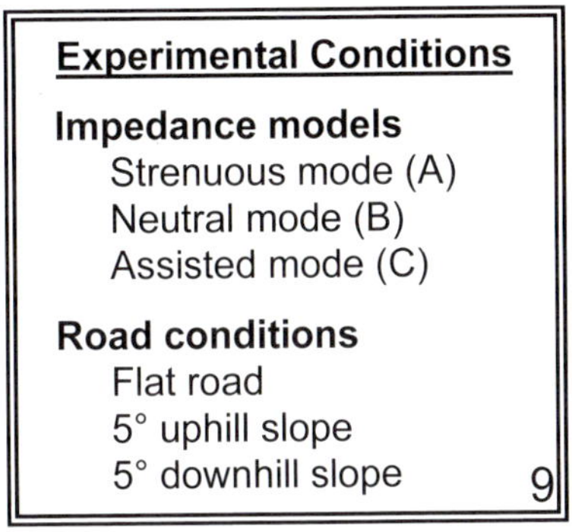

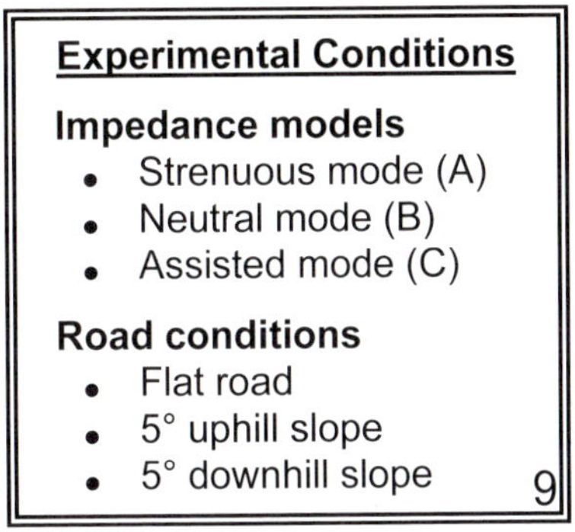

レーザーポインター

レーザーポインターで指す基本的な技術は以下の二つがある。すなわち、丸を描くことと下線を描くことである。指し終えたら、必ずレーザーポインターをオフにしなければいけない。もしもオンのままにして聴衆を見ていると、レーザードットはスライドの上にさまよう。聴衆は、発表者が何をしているかがわからなくなる。基本的に、正確に指すこと。

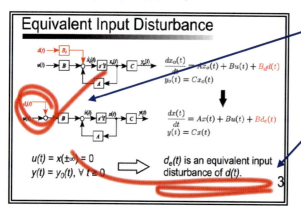

丸を描くこと
目標のまわりに1～2回レーザードットを移動する。

下線
目標の下にレーザードットを前後に移動する。

終わり

プレゼンテーションの終わりに、聴衆に聞いてもらったことに対して礼を言う。

- "Thank you very much for your attention."
- "Thank you for listening."
- "This concludes my talk. Thank you very much."

暗 記

プレゼンテーションを暗記する場合、非常によく暗記しておかなければならない。すなわち、全部のプレゼンテーションをスムーズに話すことが必要である。もし、発表者が部分的にしかプレゼンテーションを記憶していない場合、話すことを思い出そうとして何回も途中で止まったりするため、聴衆には理解しにくいブロークンイングリッシュに聞こえる可能性が大きい。

全部のプレゼンテーションを覚えることはできないならば、少なくとも最初の三、四枚のスライドの説明内容を覚えた方がよい。そうすれば、話しながら聴衆

を見ることができる。聴衆がスクリーンの中でのスライドの内容に集中した後，読み始めることができる。聴衆の多くはそれに気づかないかもしれない。

Security Blanket

Definition:
1. 精神的安定を得るために幼児がいつも手にしている毛布。
2. それがあると安心できるもの，気を落ち着かせるもの。

たとえプレゼンテーションを覚えたとしても，何かを忘れる場合に備えて，話す間に，手にテキストが印刷されたメモを持つことはよいアイデアである。それは，非常用のバックアップである。テキストはA5用紙に印刷すればよい。各スライドのテキストは，それぞれ一枚に印刷した方がよい。スライドを変えるたびに，ページをめくる。このように，見たいテキストは，すぐに見つけることができる。このメモを使わないかもしれないが，言うべきことを忘れたときは，助けになり，心強いであろう。

代わりに，Microsoft PowerPoint には，会議室のスクリーンにスライドを映しながら，それに対応するテキストをコンピュータのディスプレイで見ることができる表示モードもある。

プレゼンテーションを短くする方法

先に提出した会議論文などは，プレゼンテーションより多くの情報が含まれている。そのため，研究に関するすべての情報をプレゼンテーションで発表する必要はない。プレゼンテーションは論文の広告と見なした方がよい。すなわち，聴衆が最も面白く感じインパクトがあると思うデータと図のみをプレゼンテーションに含めればよい。目標は，研究についてすべてを説明することではなく，論文が読みたくなるほど，人々に研究に対する興味をもたらすことである。すべてを短い時間に詰め込もうとすれば，速く話さなければならなくなり，速く話すことによって，プレゼンテーションは退屈になり理解しにくくなる。プレゼンテーションで何かを説明する時間がない場合は，聴衆に論文を参照させるための文章を使用すべきである。以下にはその例文である。

- "The structure of the device is explained in more detail in the proceedings."
- "The experimental conditions are given in the proceedings."
- "If you are interested in this point, the article in the proceedings provides a fuller explanation."

質疑応答の時間

ありそうな質問リストを作成する。

質疑応答に備えるために，まず質問のリストを作成する。その質問リストは自分だけで考えるのではなく，同僚にも提案を求める。答えを書かないで，問題のみを書きとめる。練習するために，それぞれの質問を読み，大きな声でそれに英語で答える。すべての質問に英語で楽に答えることができるまで数回練習する。会議で尋ねられる質問の少なくとも半分以上がリストに載っている可能性が高い。

初めに答える。それから説明する。

それぞれの質問に対して，**説明に入る前に，最初にはっきり答える**べきである。すなわち，「イエス」か「ノー」の返事を求める質問の場合は，まず「イエス」か「ノー」をはっきりいうべきである。その後，なぜその答えをしたかを説明する。あるいは，質問者が値を求める場合，何か他のことをいう前に，最初に，その値または近似値をいった方がよい。

"I don't know."

質問への答えがわからない場合は

- "I'm sorry, I don't know."

を明確にいうべきである。その後，必要に応じて，知らない理由を説明する。

専有情報

質問者の求める情報が会社秘密であるならば，以下のように返事をする。

- "I'm sorry, but that's proprietary information."

"I don't understand."

質問を理解できない場合は，以下の文を使用できる。

- "I'm sorry, I don't understand. Could you speak more slowly, please?"
- "I'm sorry, I don't understand. Could you say that again, please?"
- "I'm sorry, I don't understand. Could you rephrase your question, please?"
  （rephrase ＝ ＜…を＞言い換える）

確　認

質問に答えたあと，質問者がその答えに満足しているかどうか疑問に思う場合もある。次の質問によって，それを確認することができる。

- "Does that answer your question?"　（「私の答えは，あなたの質問に十分答えただろうか？」）

すべての答えの後にこれをいう必要はない。

練習

自信

　良いプレゼンテーションをする鍵は，自信である。自信を得る鍵は，練習である。多数の人々の前に立つとき，誰もが緊張する。数多くのプレゼンテーションをした人でさえ，緊張する。最初の緊張をほぐすために，少し練習しすぎた方がよい。発表を行っているうちに，緊張は自然にほぐれてゆくものである。

テキストの体裁をかえる

練習をより簡単にするために，プレゼンテーションテキストの体裁を変える。
1. **原則： 単語ではなく，アイデアを話す。**人は，個々の単語ではなく，アイデアとして理解している。どんなに長くても，すべての文章は比較的短いアイデアに分割できる。そのため，**一行に一つのアイデア**という形で文章を次のページのように分解する。説明するとき，**行の途中で止まってはいけない**が，行の先頭や末尾に任意に間をおいてもかまわない。
2. テキスト文のピリオドの後に大きくて**黒い四角**をおけば，文の終わりがどこにあるかを見ることは容易である。これによって，文の終わりに適切なイントネーションを使用することを保証する。
3. プレゼンテーションの中で，どこにレーザーポインターを指した方がよいかをあらかじめ決めておく。テキストの対応する行の末尾（または先頭）に**赤丸**でマークする。

EXAMPLE: 原本

3. Equivalent Input Disturbance

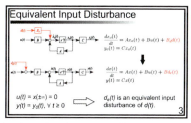

Consider this single-input single-output linear time-invariant plant. Note that the disturbance is not imposed on the control input channel. On the other hand, if we assume that the disturbance is imposed only on the control input channel, then the plant is given by these equations. Now, let's assume that the control input and the states at plus and minus infinity are all zero. If the output, y, of this plant is equal to the output, y_0, of the original plant for all t greater than or equal to zero, then we call d_e an equivalent input disturbance of the disturbance d.

EXAMPLE: 体裁をかえたテキスト

3. Equivalent Input Disturbance

Consider this single-input single-output ●
    linear time-invariant plant. ■
Note that
    the disturbance is NOT imposed
    on the control input channel. ■ ●
On the other hand,
    if we assume that the disturbance is imposed
    ONLY on the control input channel, ●
    then the plant is given by these equations. ■ ●
Now, let's assume that
    the control input
    and the states at plus and minus infinity
    are all zero. ■ ●
If the output, y, of this plant
    is equal to the output, y_0, of the original plant ●
    for all t greater than or equal to zero,
    then we call d_e
    an equivalent input disturbance
    of the disturbance d. ■ ●

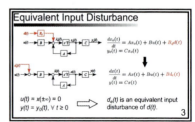

声の録音

　プレゼンテーションを練習して，話し方についてかなり自信がついたら，自分の声を録音してみよう。全部のプレゼンテーションを録音する必要はなく，最初の三，四枚のスライド分で十分である。会議と同じように，立ったままで話した方がよい。さらに，プロジェクタでスライドを映し出し，レーザーポインターを使用することを薦める。それが不可能ならば，コンピュータ画面または印刷されたスライドの紙を指す。
　規則：自分のプレゼンテーションの話し方について，一番いい批判者は自分である。 録音が完了したら，自分の声を聞く。それがよさそうに聞こえるならば，会議の大部分の人々もそれがOKに聞こえるはずである。一方，それがあまりよくないと思えば，会議の誰もがその話し方は理解しにくいと思う。自分が満足するまで，練習すること。

単語の発音

　単語の発音がわからない場合は，それをチェックした方がよい。インターネットを使用して簡単にできる。例えば，「adiabatic」の発音を知りたい場合は，単にGoogleの検索ボックスに「pronunciation adiabatic」を入力する。ネイティブ・スピーカーの発音を聞くことができる多くのウェブサイトが見つかる。

セッションの司会を務める方法

基本的な注意事項

　会議のセッション司会者を務めるように頼まれることは，大変名誉なことである。議長には，二つのおもな責任がある。
- プレゼンテーションを管理すること。
- 質疑応答を管理すること。

遅れるのは厳禁である。前もってセッションの時間と場所をチェックすることは重要である。また，できれば，あらかじめ会議室を訪ね，道順を確認する。

セッション開始前に

　セッション開始前に，次のことを行う必要がある。

◆ **共同司会者と話すこと。**
　共同司会者と相談して，役割分担を決めた方がよい。すなわち，どのように責任を分担して，どちらが何の役割を担当するかを話す必要がある。通常，一人の司会者はプレゼンテーションの件数の半分を担当する。

◆ **アシスタントが誰であるかを確認すること。**
　質疑応答の間に，専属のアシスタントがいなければ，質問とコメントをする視聴者にマイクを持っていく担当者を決める。

◆ **発表者がどのようにプレゼンテーションするつもりかを確認すること。**
　発表者は主催者の提供するコンピュータを使用する場合，セッションが始まる前に，必要なファイルをコンピュータにコピーするように頼む。一方，発表者が自分のコンピュータを使用する場合は，その発表者の順番の前に準備ができているように注意する。

◆ **いくつかの緊急質問を用意すること。**
　前もってセッションの会議論文に目を通し，それぞれにつき，一つか二つの質問を用意した方がよい。このようにして，プレゼンテーションが終わったとき，誰も質問やコメントをしない状況を避けることができる。

聴衆の会場への入室

セッションの直前に，人々は通常室外の廊下で話している。聴衆にセッションが始まろうとしていることを知らせ，入室を促す必要がある。

Example:
- "Excuse me, everyone. The session will begin in a couple of minutes. So, could you please enter the conference room and take your seats?"

セッションの開始

Example:
- "Good morning, ladies and gentlemen. Welcome to this session. My name is Richard Feynman and I will serve as the chairperson. First of all, I would like to introduce the co-chair (the vice-chairperson) Professor Yasuhiro Takahashi. He is a professor at the Tokyo Institute of Technology in Japan. We have five papers in this session. Twenty minutes are allowed for each: 15 minutes for the presentation itself and 5 minutes for discussion."

発表者の紹介

発表者の名前と所属，およびその論文のタイトルを告げることによって，発表者を紹介する。

Ex. 1:
- "The first/next/last paper is entitled *A New Avalanche Photodiode for 400G Telecom Systems*. It will be presented by Professor Hidetaka Watanabe of the Tokyo University of Technology. Professor Watanabe, please."

Ex. 2:
- "The first/next/last speaker is Professor Watanabe from the School of Information Science and Engineering of the Tokyo University of Technology in Japan. The title of the paper is *A New Avalanche Photodiode for 400G Telecom Systems*. Professor Watanabe, please."

時間制限の実施

質疑応答に十分な時間を残すように，発表者は割り当てられた時間内に終了しなければならない。時間制限を守るために，司会者としての権限を使用することが非常に重要である――それは司会者の仕事である。発表者が時間オーバーになりそうな場合，それを発表者に伝える必要がある。例えば（下記の例

のような)大きな文字で書いてあるスリップを発表者に見せること,または他の方法を使ってもいい。

2 min. left.	Your time is almost up.
Time to finish up.	Finish up, please.

プレゼンテーションの後

プレゼンテーション終了後,簡単に発表者に感謝する。

Example:
- "Thank you, Dr. Tanaka."

その後,拍手すべきことを聴衆に示すために,自分で丁寧に手を叩き始めた方がよい。

質疑応答の時間

発表者に感謝した後,質疑応答を開始する。質疑応答時間は,プレゼンテーションの重要な部分である。これにより発表者と聴衆の両方にとって,セッションの価値がより増す。そのため,上記のように,聴衆の質問がない場合にそなえ,各プレゼンテーションに質問を用意しておいた方がよい。しかし,質問が多すぎる場合は,次のプレゼンテーションが時間通りに始まるように,時間になりしだい,必ず議論をとめること。

Example:
- 聴衆に: "Are there any questions or comments?
- 手を挙げている人に: "Yes, please."

未発表者(ノーショー)の取り扱い

ノーショーがいる場合は,次の二つの方法のいずれかで,そのプレゼンテーションに割り当てられた時間を取り扱うことができる。
- 休憩を取る。
- ここまでのプレゼンテーションを議論する。

人によってある特定の論文だけを聞くためにセッションの途中で参加することもあるので,スケジュールを前倒しして変更してはいけない。

セッションの終了

Example:
- "I think it's time to close this session. Thank you very much for coming."

References

1) Nicholas J. Higham: Handbook of Writing for the Mathematical Sciences, Society for Industrial and Applied Mathematics (1998)

2) John A. Borgan: Clear Technical Writing, McGraw-Hill (1973)

3) Shogakukan Progressive English-Japanese Dictionary, 2nd ed., Shogakukan (1987)

4) William C. Paxson: The New American Guide To Punctuation, Penguin (1986)

5) American Heritage Dictionary, 2nd college ed., Houghton Mifflin (1982)

6) L. G. Alexander: Longman English Grammar, Longman (1988)

Answers

Section 1

in this work vs. in this paper

1. describe, discuss, explain, present, report (on), etc.
2. developed, devised, designed, fabricated, etc.
3. describes, discusses, explains, presents, reports (on), etc.
4. was developed, was devised, was designed, was fabricated, etc.
5. create, develop, design, devise, fabricate, invent, investigate, etc.
6. describe, discuss, explain, present, report (on), etc.

realize

1. build, construct, fabricate, make a device
2. achieve, obtain, provide, yield a large bandwidth
3. implement a function
4. build, construct, fabricate, make a component
5. build, construct, fabricate, make a digital-to-analog converter
6. carry out, perform planarization
7. achieve (目標として), obtain, provide, yield a large bandwidth-efficiency product
8. carry out, perform multiplexing
9. build, construct, make a system
10. fabricate, make an integrated circuit
11. carry out, perform control
12. carry out, perform annealing
13. fabricate, make an exclusive-or gate
14. obtain, provide, yield accurate results
15. carry out, perform a numerical analysis

Adjective Clause: Short Form

1. …results similar to those for GaMnAs.
2. The optical power coming from the two ports is 8% of …
3. …transistors with an emitter 0.6 μm wide and 3 μm long was examined.
4. A mask structure consisting of thin and thick layers was used …
5. The power of the light passing through the fibers is monitored.
6. The voltage applied to the device varied…
7. A piece of concrete containing a 0.35-mm-wide crack was used…
8. The current induced by the signal flows…
9. The angle of rotation depends on the voltage applied to the electrode of the mirror.
10. A receiver operating at a frequency of 250 GHz was fabricated and tested.
11. A sensor mounted on the tip of an optical fiber was used for the measurements.
12. The carriers injected into the device attenuate the optical power reaching the output fiber.

Changes & Differences

1. There is a 10-dB difference **in** noise level **between** these two places.
2. The use of flip-chip interconnects yields a reduction **in** crosstalk **of** about 5 dB.
3. An increase **in** the amount of moisture in dry soil can lead to a great increase **in** thermal conductivity.
4. There is a difference **in** output power **of** less than 1.0 dB **between** channels.
5. We obtained a difference **in** matching rate **of** 0.05.
6. Changes **in** the speed of rotation of a wind turbine cause fluctuations **in** the frequency and voltage of the output.
7. The refractive-index difference **between** the core and the cladding is 0.4%.
8. A reduction **in** onset gain from 4.8 to 2.8 is equivalent to an increase **in** the high-power tolerance **of** over 2 dB.
9. The results were obtained for power-supply fluctuations **of** 5%.
10. This difference **in** characteristics **between** the two samples could be due to a difference **in** current-blocking properties.
11. A 20% drop **in** the peak-to-peak voltage was simulated.
12. A 3-dB improvement **in** the signal-to-noise ratio and a 3-dB reduction **in** noise were obtained.
13. The difference **between** the free energies of Reactions A and B is very small.
14. There is a small difference **in** free energy **between** Reactions A and B.

operating principle

1. operat**ion**
2. operat**ing**
3. operat**ion**
4. operat**ing**
5. operat**ion**
6. operat**ing**
7. operat**ing**
8. operat**ion**

evaluate vs. estimate

PRACTICE A

1. evaluate
2. estimate
3. estimate
4. estimate
5. evaluate
6. evaluate
7. evaluate
8. estimate
9. estimate
10. evaluate 　[「Performance」は, 特定の変数 (例えば, 速度) であることをすでに論文で定義されている場合, 「performance」をestimateすることが可能である。]
11. estimate

PRACTICE B

1. determined, estimated
 　[「Measured」も, 特定の種類

Answers

の顕微鏡にとって可能である。]
2. identified
3. determined
4. determine, examine, investigate
5. determined, examined, investigated
6. observe, examine, investigate
7. measured, examined, investigated

Dynamic Verbs 1

1. Hot electrons **were injected into** the transistors.
2. The electrical interface **limits** the bandwidth.
3. Figure 3 **illustrates** the experimental setup.
4. The polymer layer **was removed**.
5. This paper **is organized** as follows:
6. The displacement **can be accurately measured** with an encoder.
7. It is difficult **to dramatically improve** the performance.
8. Voltage drops in a utility grid **can severely damage** sensitive loads.
9. The wavelength **was converted** from λ_1 to λ_2.
10. K_P and $F(s)$ **can be designed** independently.

Prepositions 1

1. ×
2. ×
3. to
4. as
5. to
6. of
7. ×
8. ×
9. ×
10. on

Section 2

propose

1. created, developed, designed, devised, etc.
2. the, our new, our newly developed, a fabricated, etc.
3. created, developed, designed, devised, etc.
4. describes, explains, presents, reports on, etc.
5. created, developed, designed, devised, fabricated, invented, etc.
6. create, develop, design, devise, etc.

Lists

1. This modulator provides **high speed, a low driving voltage, and other required characteristics**.
2. **Neural networks, adaptive fuzzy control, and other intelligent control methods** have been used to deal with this problem.
3. The acceptable crack width depends on **humidity and other ambient conditions**.
4. The components are affected in different ways by **aging, voltage fluctuations, variations in temperature, and other factors**.
5. The report covers **propagation problems, system design parameters, possible applications, and other technical and operational issues**.

6. **A dead zone, hysteresis, and other nonlinearities** can seriously degrade control performance.
7. There is little space for **an isolator, filters, and other optical devices**.

contain vs. include

1. contains
2. including (Meaning 4)
3. included (Meaning 3)
4. contains, includes (Meaning 2)
5. included (Meaning 3)
6. contain
7. include (Meaning 1)
8. includes (Meaning 4)
9. contain, include (Meaning 2)
10. including (Meaning 4)
11. include (Meaning 1)
12. including (Meaning 1)
13. contain
14. contains
15. include (Meaning 4)
16. includes (Meaning 1)
17. includes (Meaning 2)

in case of fire

1. The values are 8.4 mW **for Device A** and 11.2 mW **for Device B**.
2. This phenomenon appears only **when the input power is greater than a certain value**.
3. This figure shows pulse patterns **for an output voltage of 200 mV**.
4. **When the voltage is low**, light passes through.
5. These figures show the performance **for** **156-Mb/s signals**.
6. The resistivity increased monotonically with annealing time **when the zirconium layer was 200 nm thick**.
7. **When the total thickness is greater than 1 μm**, the coupling efficiency increases dramatically.
8. Chemical polishing reduces the threshold voltage **for long-channel devices**.
9. **When a defect is completely within the silicon**, it does not give rise to a defect in the gate oxide.
10. **For a conventional DFB laser**, the lasing mode is very stable.

Connecting Nouns

1. **the length of** a/the Si-Si bond
2. **the change in** the critical dimension
3. **the calculation of** the intensity profile
4. **the scanning frequency of** the mirror
5. **the thickness of** the GaAs buffer layer
6. **the replication of** sub-100-nm space patterns
7. **the increase in** the amount **of** data traffic
8. at **coding rates of** less than 10 Mb/s
9. **a method of** controlling **the amount of** surface acid

Answers

Hyphen (-)

1. We made 3-µm-long devices.
2. OK
3. We made 3-micron-long devices.
4. OK
5. We made a device 3 µm long.
6. OK
7. a 15-minute presentation
8. OK
9. a low-voltage LSI
10. OK
11. a gate width of 5 µm
12. at a low V_{th}
13. OK
14. a high-temperature process
15. a one-dimensional system
16. a GaAs buffer layer 480 nm thick
17. a 0.5-µm-thick layer of resist
18. lattice-matched substrates
19. Each device is 2.5 mm long.
20. a high operating frequency of 200 MHz

Dynamic Verbs 2

1. The calculations **take** these atoms into account.
2. Wire bonding **generates** mechanical stress.
3. A lens **focused** the light on the device.
4. The load capacitance **suppresses** voltage oscillations.
5. The feedback dramatically **improves** the characteristics.
6. The light **raises** the temperature.
7. This electrical excitation **generates** ballistic electrons.
8. This model **includes** the gate-drain capacitance.
9. This circuit **reduces** the number of connections by 75%.
10. In Section 4, numerical examples **demonstrate** the effectiveness of the method.

Prepositions 2

REVIEW

a. to c. × e. as
b. ×   d. ×

CHECK YOUR KNOWLEDGE

1. of   5. to   9. for
2. on   6. with 10. of
3. ×   7. ×
4. to   8. on

Section 3

compared to vs. than

1. This method is **much more effective than** conventional ones.
2. The processing power is **much lower than** that of conventional machines.
3. Holes have **a larger effective mass than** electrons <u>do</u>.
4. The gains are **much higher than** this value.
5. The new circuit **is 25% smaller than** a conventional one. / The size of the new circuit **is 25% smaller than** that of a conventional one.
6. The accuracy is **lower than** that of the simulation results.

for –ing

1. OK［「be used for」＝熟語］
2. The layer structure was designed **to achieve** a fast response.
3. OK
4. A thermoelectric cooler was added **to control** the temperature of the chip.
5. **To track** moving objects, it is necessary to match key features in neighboring frames.
6. OK［「be needed for」＝熟語］
7. OK
8. Experiments were performed **to verify** the proper operation of the device.
9. We optimized the laser **to improve** the performance of the laser array.
10. **To measure** the resistance, the cathode was attached to a Pt mesh.
11. OK
12. OK［「be used for」＝熟語］
13. The composition of the WSiN was measured **to determine** the effect of an RF bias.
14. OK

can could

1. The good agreement between the experimental and simulation results **demonstrates** that the simulation method is correct.
2. OK［可能性の限界を示す。］
3. OK［可能性］
4. The resolution **was estimated** to be 0.19 mm from the dimensions of the device.
5. This fabrication technique **prevents** the unintentional formation of parasitic islands.
6. OK［条件によって(例えば十分厚いなら)、可能性がある。］
7. There **are** three possibilities: (a) no diode switches, (b) one diode switches, and (c) two diodes switch.
8. OK［可能性］
9. OK［(条件によって(例えば十分厚いなら)、可能性がある。しかし、「Silicon **blocks** high-energy X-rays.」でもよい。］
10. These results demonstrate that our algorithm **improves** the verification accuracy.
11. OK［可能性。いろいろな分け方がある。］
12. We **obtained** an output power of 1.7 W.

as a result

1. The actuation voltage was swept up from 0 to 20 V and then down to 0 V. **The results showed that** the switch turned on at a voltage of 10 V and turned off at 9 V.
2. OK
3. **BET measurements showed that** the specific surface areas of the oxides have the values listed in Table 3.
4. OK

5. Section 3 describes a high-speed optical-coherence-tomography (OCT) system. Section 4 presents OCT images made with the system.
6. OK
7. Figure 7 shows XRD patterns of $Mn_{2-x}Fe_xO_3$ heated to 500°C. When x was 0, 0.2, or 0.4, a solid solution of Mn_2O_3 formed. /OR/ XRD patterns of $Mn_{2-x}Fe_xO_3$ heated to 500°C (Fig. 7) show that, when x was 0, 0.2, or 0.4, a solid solution of Mn_2O_3 formed.
8. OK
9. OK

becomes vs. is
PART A
1. OK
2. **As V_{DD} becomes smaller**, the delay increases. / When V_{DD} is small, **the delay is large**.
3. If X is large enough, Z **is** negligible.
4. OK

PART B
1. OK
2. With this design, the gain of one stage **is** G_{total}/n, where G_{total} is the total gain of the amplifier and n is the number of stages.
3. OK
4. To track moving objects, it **is** necessary to match feature points in neighboring frames.
5. OK

Unnecessary Repetition
1. We paid special attention to the influence of the subband system, and to the possibility of observing **it** at high temperatures.
2. The overcoat was about 0.1 μm thick. **It** was removed after baking.
3. This mesa structure is not easy to bury because **it** has no mask on top.
4. This is an example of an action logic table. **It** can also be transformed into Prolog.
5. We developed a global router for high-speed bipolar LSIs. **It** minimizes areas....
6. We observed the spots to estimate the amount of relaxation in the lattice parameters. **The results revealed** that **they** are relaxed in two steps.
7. This is the circuit we designed. **It** was fabricated on a CMOS process.

Prepositions 3
REVIEW
a. to   e. for   i. ×
b. on   f. on   j. of
c. ×   g. ×
d. with   h. on

CHECK YOUR KNOWLEDGE
1. for   5. in   9. on, of
2. to   6. of   10. to
3. ×   7. with
4. On   8. from

Section 4

respectively

1. Wrong
2. Wrong
3. OK
4. OK
5. Wrong
6. OK
7. Wrong
8. OK
9. OK
10. Wrong
11. OK

common vs. popular

1. common, popular
2. common
3. popular
4. common
5. popular
6. common, popular
7. common

recently

1. OK
2. These systems **are now being investigated**...
3. OK
4. Recently, the operating speed of CMOS LSIs **reached** about 4 GHz.
5. OK
6. A great deal of attention **is now being paid** to these defects.
7. 通常、「recently」は特定な時間と一緒に使わない。
   Recently, e-beam lithography was used to fabricate this circuit. / E-beam lithography was used to fabricate this circuit <u>in 2014</u>.
8. OK
9. **Nowadays**, portable equipment uses low-voltage LSIs.
10. This method **is now** widely used.
11. OK
12. OK
13. It **is now** becoming more important to reduce costs.

Adjective Formation (-ing)

PART A
1. It is a **wafer-holding** mechanism
2. It is **zirconium-containing** copper.
3. It is **industry-leading** performance.

PART B
1. It is fiber that **reduces dispersion**.
2. They are techniques that **reduce noise**.
3. It is equipment that **bonds wires**.

Adjective Formation (-ed)

PRACTICE A

PART A
1. It is **computer-aided** design.
2. It is **boron-doped** silicon.
3. It is a **laser-generated** pulse.

PART B
1. They are shoes that **are made by** hand.
2. They are components that **are based on** semiconductors.
3. It is a technique that **is oriented toward** speed.

Answers 193

PRACTICE B

1. It is a **water-rinsed** resist.
2. It is a **packet-switching** network.
3. It is a **liquid-filled** section.
4. It is **time-averaged** power.
5. It is a **current-blocking** structure.
6. It is **letter-sorting** equipment.
7. It is a **low-temperature-grown** device.
8. They are **user-specified** parameters.
9. It is a **rate-limiting** factor.

Comma 1

1. Unlike radio waves, terahertz waves are not scattered very much by dust, soot, or smoke.
2. The sensor unit weighs only 5 g, so it is light enough to be unnoticeable by a wearer.
3. Table 2 compares length, area, and execution time.
4. Direct bonding is an easy way to put a laser on Si, but there are two drawbacks.
5. Consider, as an example, an experiment in which a coin is tossed three times.
6. The use of passive couplers provides a small phase error, but it makes the loss large.
7. All cases used 0.75-μm-thick PMMA resist, with the results being an average for isolated lines, isolated spaces, and line-space arrays.
8. Then, it gradually becomes paler and finally turns grayish.
9. The n-type dopant is silicon, and the p-type is carbon.
10. On the other hand, if X is low, Y will be low after the clock goes high.
11. However, if these constraints are present, problems arise if one attempts to simplify the timing graph.
12. The overhead is 4 cycles, so the total number of cycles is 68.
13. If we use a wavelength filter as a multiplexer, the loss is low, but the size is large.
14. With image placement targets as low as 35 nm, all contributions must be minimized, including that from the e-beam system and process-induced distortion.
15. Common layout procedures, such as symbolic layout and layout compaction, destroy the symmetry of critical analog layouts, impacting performance.
16. When the clock input is low, transistors P_3 and P_4 act as resistive loads for the first stage, which acts as a linear amplifier for small input swings and as a swing limiter for large input swings.
17. High voltage reduces scattering, resulting in better resolution, straighter side walls, and smaller proximity effects.

Unnecessary Words 1

1. **The shielding** provided by the 3D structure enables a variety of useful devices to be fabricated.
2. To monolithically integrate the devices, we employ **regrowth**.
3. OK [The capital letter 'S' in "*Stark*" indicates that it is a person's name (i.e., Johannes Stark).]
4. **Anisotropic Si etching** is one of the most important technologies for **bulk micromachining**.
5. **The collimation** is caused by the flaring of the potential boundary.
6. OK [A short channel is desirable, but shortening the channel has undesirable side effects.]
7. This structure is essential for achieving **a low threshold current and a high output power**.
8. **Offset canceling** is employed in the limiting amplifier.
9. Two special techniques were used for the fabrication: **image reversal** and **oxidation**.
10. High speed results from the suppression of **current blocking**.
11. OK [An air bridge is a physical structure. An air-bridge technique is a method of making that structure.]
12. The p+ regions are formed by Zn diffusion during **alloying**.
13. This device employs **a Coulomb blockade** to manipulate individual electrons.
14. OK [The capital letter 'A' in "*Auger*" indicates that it is a person's name (i.e., Pierre Victor Auger).]

Prepositions 4

REVIEW

a. for e. on i. on, of
b. to   f. ×   j. ×
c. ×   g. of
d. for h. of

CHECK YOUR KNOWLEDGE

1. with   5. to   9. between
2. to   6. in   10. ×
3. to   7. to
4. into   8. on/about

Section 5

has been used vs. is used

1. OK
2. A lot of multimode fiber **is being used** for LANs in office buildings.
3. OK
4. Three main methods **are used** to grow crystals of organic materials.
5. OK
6. OK
7. GaAs FETs **are widely used** in circuits operating at microwave frequencies.
8. OK

9. The amplifier we developed **is being used** in the transmitter and receiver of a 120-GHz-band wireless system.
10. OK
11. Conventionally, lead-acid batteries **are used** for backup power supplies.

by vs. with

1. with
2. with
3. with
4. by
5. with
6. with
7. with
8. by
9. by, by
10. with
11. with

Keep Related Words Together

1. a method of **estimating the current** / an estimation method **for** the current
2. electrons **injected into** the base
3. the waveform of signals **transmitted over** our new microstrip line
4. We have developed a new technique for **trimming planar lightwave circuits**.
5. Data **stored in a 1-V SRAM** is destroyed at a voltage of 1.8 V.
6. to develop a method of **designing DC power networks** / to develop a design method **for** DC power networks
7. a technique for **integrating transistors and photodiodes** / an integration technique **for** transistors and photodiodes
8. a pattern **replicated by ECR plasma etching**
9. The bias voltage **applied to the device** was –1.0 V.
10. the problem of **estimating the normalized longitudinal force**

multi-

1. **multi-electrode** device
2. **multichip** module
3. **multifunction/multifunctional** circuit
4. **multilayer/multilayered** structure
5. **multiprocessor** system

fixed

1. In our model, the thickness of a metal line **is** 0.5 μm.
2. OK
3. In this experiment, the received power was **set to** –20 dBm.
4. The heat flow profile inside the device changes, even though the total power dissipation is **constant**.
5. OK
6. The shielding protects the fibers **attached to** the plug.
7. In the simulations, the input frequency **was** 2 MHz.
   (OR ...**was a constant** 2 MHz.)
8. The number of "1" bits in each packet was **constant**.
9. The frequency **was** 2 GHz.
10. Four fiber-lens assemblies are **attached to** the metal block.
11. The timing conditions for the

clock and data are **set** inside the chips.
12. The current injected into the active region **was** 120 mA.
(OR ...**was a constant** 120 mA.)

Comma 2

1. However, the photovoltaic parameters are very low. For example, the fill factor is only 44%, and the conversion efficiency is just 0.07%.
2. It is difficult to estimate the spring constant, κ, and the damping factor, γ.
3. Mori University, Building 65, W-Wing 804C, 5-2-9 Okubo, Shinjuku, Tokyo 170-8432, Japan
4. The wavelength was 203 nm; the intensity, ~110 mJ/cm^2/pulse; and the repetition rate, 16 pulses/s.
5. One source is the variation in the neutral current, which is caused, for example, by an unbalanced load.
6. One plate is soldered to Package 1; and the other, to Package 2.
7. The junction temperature, T_j, was about 225°C.
8. This study employed the LQR optimal control method (see, for example, [14]).
9. The transmitter consumes 1.2 W; and the receiver, 0.75 W.
10. For example, if the key "2222" is input, the flag for address "2222" is read.
11. To tune the output frequency, we can change either the reference frequency, f_{comb}, or the multiplication order, N.
12. Linkers Inc., 6F Sato Building, 4-7-6 Ginza, Chuo, Tokyo 106-0328, Japan
13. When the emitter-base voltage, V_{BE}, was low, the base current, I_B, increased significantly with time.

Unnecessary Words 2

1. The first test **clarified** the degradation modes.
2. Optical fibers **transport** light to and from the probe.
3. A Teflon lens **focused** the signal on the detector.
4. This terraced structure **improves** the heat dissipation of the chip.
5. An electro-optic probe **detected** reflected signals.
6. The tungsten **reduces** the resistance of the electrodes.
7. Etching with HF vapor **removes** the SiO$_2$ sacrificial layer.
8. These chemical agents **improve** the adhesion of polymer to a substrate.

Prepositions 5
REVIEW

a. of   e. between   i. with
b. to   f. to   j. to
c. to   g. into
d. with h. in

CHECK YOUR KNOWLEDGE

1. with   5. in   9. in
2. to   6. on/about   10. In
3. to   7. to
4. of   8. of

Section 6

know vs. find out

1. find out
2. know
3. find out
4. find out/know
5. find out
6. find out
7. know
8. find out
9. know
10. find out

none, one, some, most, all

1. Table 1 lists **some of the** specifications of the new modulator.
2. Phase noise is **a** big problem.
3. The process is complex, and **many of the** process factors are coupled.
4. **One of the** variables in the formula is maximum heart rate.
5. There are **many** concrete structures in the world.
6. **All (of) the** components are commercially available.
7. Section 5 presents **some** experimental results.
8. This remote-driving scheme avoids **many of the** problems associated with contact-based schemes.
9. Since no surface waves are generated, **most of the** power goes into the waveguide.
10. **All** telecommunications devices require a low loss and a large dynamic range.
11. Terahertz waves can penetrate **most** non-metallic materials.
12. Millimeter waves have **some** interesting applications.
13. **All (of) the** measurements were carried out at a temperature of 40°C.
14. **Most** high-speed digital-to-analog converters are based on this architecture.
15. **All** possible pairs of the items are formed, and a search is carried out on each pair.
16. **Some of the** unit cells are arrayed two dimensionally.
17. N_2O has **an** absorption line at 16.6 μm.
18. PECST is **a** Japanese national project.
19. **Most of the** light injected into the Ge is absorbed.
20. A mobility of 22,000 cm^2/Vs is **one of the** highest values ever reported.
21. This new device overcomes **some of the** imitations of conventional devices.
22. **Most of the** frequency components are below 10 Hz.
23. **One of the** electrodes of our battery consists of MnO_2.
24. Assumption 2 holds in **many** real-world systems.

Meaningless –ing

1. **After the annealing time was shortened**, the blue shift disappeared.
2. **When these functions are combined**, various types of processing can be performed.
3. These structures **provide** a high mobility. / These structures **enable** a high mobility **to be obtained**. / A

198   Answers

high mobility **was obtained with** these structures. / These structures **were used to obtain** a high mobility.
4. **An analysis of the results revealed** two patterns. / **An analysis of the results showed that** there were two patterns.
5. **Based on** this property, interesting imaging applications have recently been developed. / This property **has recently been used to develop** interesting imaging applications.

issue vs. problem

1. problem
2. issue
3. issue
4. problem
5. problem
6. issue
7. issue
8. issue
9. problem
10. problem*
11. issue

* *How to do something is usually a problem.*

problem with/of

1. with
2. with
3. of
4. with
5. with
6. of
7. of

Semicolon (;)

1. The open circles are for the pile-up model; and as you can see, the agreement is excellent.
2. This figure illustrates a number of key characteristics of the kink: The kink in I_D occurs approximately at a constant V_{DG} of 1.2 V; the size of the kink appears to increase with increasing V_{GS}; and the onset of the kink coincides with the appearance of I_{SG} and with a prominent rise in E_G, presumably due to hole collection by the gate.
3. Below the line, diffusion is dominant; and above, drift is dominant.
4. Unlike digital LSIs, most of the area of MMICs is occupied by passive elements, such as transmission lines, inductors, and capacitors; and reducing their size is the best way to miniaturize MMICs.
5. Circuit extraction is an important step in VLSI circuit design verification; it provides the link between the physical design and its verification phases.
6. As shown in Fig.8, when the phase is 160°, the best-focus position does not shift at all; and the depth of focus is as wide as that without spherical aberration.
7. The data timing is ideally centered at zero and tracks the bit length; it is ideally ±0.75 ns at a bit length of 1.5 ns, and ±1.5 ns at a bit length of 3.0 ns.

(Fig. 3)

1. If the receiver is near a conductive wall **(Fig. 2)**, the wall changes the pattern of the quasi-electrostatic field.
2. A TORA **(Fig. 6)** consists of a cart and an eccentric rotational proof mass.
3. The simulation results **(Fig. 5)** show that the settling time was less than 13 s.
4. The control system **(Fig. 1)** consists of four parts.
5. The relationship between β and J_1 **(Fig. 5)** shows that J_1 is small when $0.25 < \beta < 0.55$. / From the relationship between β and J_1 **(Fig. 5)**, it is clear that J_1 is small when $0.25 < \beta < 0.55$.
6. Simulation results **(Fig. 7)** for the two parameter sets show that the control performance is better for Set B than for Set A.
7. The model **(Fig. 5)** has 13 nodes.

Prepositions 6
REVIEW
a. to
b. to
c. ×
d. to
e. of
f. with
g. for
h. to
i. to
j. ×

CHECK YOUR KNOWLEDGE
1. from
2. in
3. of
4. into
5. into
6. with
7. for
8. with
9. into
10. on

Section 7

Bad Passives

1. The variation **originates** in the fabrication process.
2. A thin layer of SiO_2 **remained** on top.
3. The GaAs layer may **disappear** during etching if...
4. Limit-cycle oscillations **occur**.
5. It is impossible to avoid the bandwidth limitation **originating** from the carrier response of the semiconductor.
6. When a bit error **occurs**, the device requests that the data be resent.
7. The data in the selected memory cells **appear** on the bit lines.
8. This high value suggests that some hydrogen still **remains**.

composition vs. content

1. content
2. composition
3. content
4. content
5. composition

maintain vs. remain

1. The output power **remains constant**.
2. The extinction ratio **remains over 13 dB**.
3. The Ni and Ti contents **remain the same** during the ion-exchange process.

4. The optical quality of the output light **should remain constant** (*OR* **the same**), even after thousands of circulations.
5. The energy profile around the edge **remains steep**.
6. The loss **remained sufficiently low**.

recover vs. restore

1. recovered
2. restores
3. recovers
4. restores
5. recover

Larger For A Than For B

1. The control performance **was better for** set A **than for** set B.
2. The etching rate **is higher for** the In component **than for** the Ga component.
3. The attenuation of the received power **was larger during** snow **than during** rain.
4. The peak intensity **is much larger for** the quantum wells **than for** the GaN layer.
5. The currents **are much smaller in** an adiabatic SRAM **than in** a conventional SRAM.
6. The multiplication factor **is larger at** long wavelengths **than at** short wavelengths.
7. The stability condition **is more conservative for** Theorem 3 **than for** Lemma 2.
8. The growth rate **was smaller for** the carbon-doped sample **than for** the undoped sample.
9. The allowable phase error **is smaller for** a 20-Gb/s demodulator **than for** a 10-Gb/s one.
10. The frequency characteristics **are flatter for** an optical modulator **than for** a millimeter-wave mixer.
11. The requirements **are stricter for** interchip interconnections **than for** intrachip interconnections.

Prepositions 7
REVIEW

a. into
b. from
c. with
d. of
e. of
f. with
g. ×
h. from
i. ×
j. on

CHECK YOUR KNOWLEDGE

1. to
2. to
3. about
4. on
5. on
6. In
7. in
8. in, with
9. in
10. about

Section 8
almost

1. **Most/Almost all** DC networks contain a DC-DC converter.
2. OK
3. OK
4. The temperature of a coke oven is **usually/mainly** measured manually.
5. The operating speed is

mainly/largely determined by the peak current density.
6. OK
7. The last stage of the amplifier **outputs almost** 1 V.
8. OK

whose

PRACTICE A
1. a beam **with a diameter of** about 6 nm
2. a signal **with a wavelength, λ, of** 1,300 nm
3. pulses **with a width of** less than 30 ps
4. a layer **with a thickness of** 15 nm
5. an antenna **with a gain, A_a, of** 48.7 dBi

PRACTICE B
1. signals **with frequencies that** extend down to several kilohertz
2. a toothed antenna **with teeth that** correspond to frequencies from 150 GHz to 2.4 THz
3. a converter **with capacitors that** are fabricated on the chip
4. a package **with a coefficient of thermal expansion that** is close to that of $LiNbO_3$
5. an antenna **with a length that** is half the wavelength of the carrier signal
6. foam cubes **with edges that** are 25 cm long

PRACTICE C
1. fixed wireless access, **which has a top speed of** 622 Mb/s,
2. The image rejection ratio is 49 dB, **which satisfies** the specifications for short-range wireless systems.
3. the module, **which is the same size as** a standard LD module,
4. We used Kovar, **which has characteristics that** are close to those of the glass.

complete(ly) vs. perfect(ly)

1. perfectly
2. completely
3. completely
4. perfect
5. completely
6. perfect
7. completely
8. completely

Prepositions 8

REVIEW
a. as   e. about   i. ×
b. in   f. on, of   j. into
c. on/about   g. to
d. on   h. for

CHECK YOUR KNOWLEDGE
1. at   5. in   9. to
2. at   6. to
3. on   7. between
4. on   8. in, between

Index

【A】
a/an	116
Abbreviation vs. Symbol	21
Adjective Clause: Short Form	11
Adjective Formation (-ed)	84
Adjective Formation (-ing)	83
adopt	40
against	156
all	118
all (of) the	118
almost	152
and so on	31
another	136
apply	46
approach	58
approach to	114
as	67
as a result	62
assess	8
at first	17
ave	3
avg	3

【B】
Bad Passives	133
be consistent with	142
be dependent on	35
be expected	57
because	134
becomes	67
by	98
by contrast (to)	39

【C】
calculated vs. calculation	132
can	59
can not	41
can't	41
cannot	41
Capitals	168
change (変化を示す動詞)	16
change in/of	14
Changes & Differences	14
Change vs. Comparison	166
coincide	105
Colon (:)	69
Comma 1	87
Comma 2	108
commercialized	161
common	78
compare between	162
compared to	53
compensate	86
complete(ly)	157
composition	135
confirm	7
Connecting Nouns	44
consist of	61
contain	36
content	135
contribute to	132
control	74
conventional	86
correspond	106
could	59

【D】
damage	54
damages	54
Dash	168
decrease	166
demonstrate	8
depend on	35
determine	7, 8
Dynamic Verbs 1	25
Dynamic Verbs 2	50

【E】
e.g.	31
each	65
effective	92

efficient	93
enable	22
enough	82
estimate (n.)	20
estimate (v.)	19
estimation	20
etc.	31
evaluate	8, 19
examine	8
express	163

【F】

find, found	8
find out	7, 113
first	17
firstly, secondly, etc.	17
fixed	103
flow	155
fluctuations	149
for	42
for example	31
for -ing	55

【H】

has been used	94
however	150
Hyphen	48

【I】

improve	138
in case of	42
in contrast (to)	39
in this paper	2
in this work	2
include	36
increase	166
including	31
introduce	81
is	67
is expected	57
is thought	64
is used	94
issue	121

【K】

Keep Related Words Together	100
key to	114
know	113

【L】

Larger For A Than For B	146
Lists	30

【M】

maintain	137
marked	73
Meaningless -ing	120
measured vs. measurement	132
monotonic	144
monotonous	144
most	117, 118
most of (the)	117, 118
multi-	102

【N】

neither...is	142
none	118
none of (the)	118
not A or B	141
novel	3
number	96

【O】

obvious	123
on the contrary	39
one	118
one of (the)	116, 118
operating principle	18
optics is vs. optics are	124
Oral-Presentation Hints	170

【P】

Parentheses	145
perfect(ly)	157
performance	154
performances	154
popular	78
prepare	66
Prepositions 1	26

Prepositions 2	51	**[S]**	
Prepositions 3	71	saturate	139
Prepositions 4	90	seem	64
Prepositions 5	111	Semicolon (;)	127
Prepositions 6	130	should	57
Prepositions 7	147	simple	80
Prepositions 8	169	simplified	80
probably	64	simulated vs. simulation	132
problem	121	since	134
problem with/of	126	Slash	167
promising	58	small AND red?	160
Pronoun	70	so	125
property	45	so-called	123
proportion	63	some	118
proportional	63	some of (the)	118
propose	28	Space	23
proposed device	28	specialized	161
Punctuation: Capitals	168	Specifying Values	35
Punctuation: Colon (:)	69	standardized	161
Punctuation: Comma 1	87	Study	2
Punctuation: Comma 2	108	Style: (Fig. 3)	129
Punctuation: Dash	168	Style: Dynamic Verbs 1	25
Punctuation: Hyphen	48	Style: Dynamic Verbs 2	50
Punctuation: Parentheses	145	Style: Larger For A Than For B	146
Punctuation: Semicolon (;)	127	Style: Unnecessary Repetition	70
Punctuation: Slash	167	Style: Unnecessary Words 1	89
Punctuation: Space	23	Style: Unnecessary Words 2	110
[R]		such as	31
reach	159	summarize	159
realize	4	Symbol	21
recently	79	**[T]**	
recover	143	than	53
reduce	166	that	9, 11
reduction in/of	16	that-形容詞節	10
remain	137	the all	99
remarkable	73	the another	99
Reporting	2	the both	99
Research	2	the each	99
respectively	76	the other(s)	136
restore	143	the some	99

then	115, 150
therefore	125, 150
thus	150
tolerance	75
traditional	105

【U】

Units	21
Unnecessary Repetition	70
Unnecessary Words 1	89
Unnecessary Words 2	110

【V】

variation	149

【W】

when	42
which	9, 13
which-形容詞節	10
whose	153
with	98
with increasing frequency	151
without A or B	141

【X】

XXXable	164

【和語】

確認した	8
確認する	7
確認するために	8
区別情報	9
形容詞節：省略形	11
形容詞の比較形	53
研究 vs. 報告	2
懸垂分詞	120
口頭発表：I vs. We	173
口頭発表：アウトライン	172
口頭発表：暗記	176
口頭発表：終わり	176
口頭発表：形式のレベル	171
口頭発表：質疑応答の時間	178
口頭発表：スライド	174
口頭発表：聴衆を見ること	173
口頭発表：次のトピックへの移行	172
口頭発表：初めに	171
口頭発表：プレゼンテーションを短くする方法	177
口頭発表：略語の導入	173
口頭発表：レーザーポインター	176
口頭発表：練習	179
口頭発表のヒント	170
コンマ	9, 87, 108
最　近	79
実現する	4
セッションの司会を務める方法	181
それぞれ	76
代名詞	70
比較の標準	53
評価する	20
報告 vs. 研究	2
補足情報	9
名詞＋-ed形容詞の作成	84
名詞＋-ing形容詞の作成	83
リスト：一項目一行	33
リスト：インライン（行内）	30
リスト：コロン（：）	30
リスト：セミコロン（；）	34

―― 著者略歴 ――

Richard Cowell（リチャード　カウェル）
1970 年　スタンフォード大学数学科卒業
1980 年　インテック ジャパン株式会社勤務
2013 年　株式会社リンクグローバル
　　　　　ソリューション勤務
2016 年　逝去

佘　錦華（しゃ　きんか）
1993 年　東京工業大学大学院理工学研究科
　　　　　博士後期課程修了
　　　　　博士（工学）
1993 年　東京工科大学講師
2001 年　東京工科大学助教授
2007 年　東京工科大学准教授
2010 年　東京工科大学教授
　　　　　現在に至る

マスターしておきたい 技術英語の基本　―決定版―
Mastering the Basics of Technical English, Definitive Edition

© Richard Cowell, Jinhua She 2006, 2015

2006 年 6 月 28 日　初　版第 1 刷発行
2014 年 8 月 30 日　初　版第 11 刷発行
2015 年 12 月 10 日　改訂版第 1 刷発行
2022 年 1 月 15 日　改訂版第 5 刷発行

検印省略	著　者	Richard Cowell
		佘　　錦　華
	発行者	株式会社　コ ロ ナ 社
		代表者　牛来真也
	印刷所	新日本印刷株式会社
	製本所	有限会社　愛千製本所

112-0011　東京都文京区千石 4-46-10
発行所　株式会社　コ ロ ナ 社
CORONA PUBLISHING CO., LTD.
Tokyo Japan
振替 00140-8-14844・電話 (03) 3941-3131 (代)
ホームページ　https://www.coronasha.co.jp

ISBN 978-4-339-07799-5　C3050　Printed in Japan　　　（新井）

〈出版者著作権管理機構　委託出版物〉
本書の無断複製は著作権法上での例外を除き禁じられています。複製される場合は，そのつど事前に，
出版者著作権管理機構（電話 03-5244-5088，FAX 03-5244-5089，e-mail: info@jcopy.or.jp）の許諾を
得てください。

本書のコピー，スキャン，デジタル化等の無断複製・転載は著作権法上での例外を除き禁じられています。
購入者以外の第三者による本書の電子データ化及び電子書籍化は，いかなる場合も認めていません。
落丁・乱丁はお取替えいたします。

辞典・ハンドブック一覧

農業食料工学会編
農業食料工学ハンドブック B5 1108頁 本体36000円

安全工学会編
安全工学便覧(第4版) B5 1192頁 本体38000円

日本真空学会編
真空科学ハンドブック B5 590頁 本体20000円

日本シミュレーション学会編
シミュレーション辞典 A5 452頁 本体9000円

編集委員会編
新版 電気用語辞典 B6 1100頁 本体6000円

編集委員会編
改訂 電気鉄道ハンドブック B5 1024頁 本体32000円

日本音響学会編
新版 音響用語辞典 A5 500頁 本体10000円

日本音響学会編
音響キーワードブック―DVD付― A5 494頁 本体13000円

電子情報技術産業協会編
新ME機器ハンドブック B5 506頁 本体10000円

編集委員会編
機械用語辞典 B6 1016頁 本体6800円

編集委員会編
制振工学ハンドブック B5 1272頁 本体35000円

日本塑性加工学会編
塑性加工便覧―CD-ROM付― B5 1194頁 本体36000円

精密工学会編
新版 精密工作便覧 B5 1432頁 本体37000円

日本機械学会編
改訂 気液二相流技術ハンドブック A5 604頁 本体10000円

日本ロボット学会編
新版 ロボット工学ハンドブック―CD-ROM付― B5 1154頁 本体32000円

土木学会土木計画学ハンドブック編集委員会編
土木計画学ハンドブック B5 822頁 本体25000円

土木学会監修
土木用語辞典 B6 1446頁 本体8000円

日本エネルギー学会編
エネルギー便覧―資源編― B5 334頁 本体9000円

日本エネルギー学会編
エネルギー便覧―プロセス編― B5 850頁 本体23000円

日本エネルギー学会編
エネルギー・環境キーワード辞典 B6 518頁 本体8000円

フラーレン・ナノチューブ・グラフェン学会編
カーボンナノチューブ・グラフェンハンドブック B5 368頁 本体10000円

日本生物工学会編
生物工学ハンドブック B5 866頁 本体28000円

定価は本体価格+税です。
定価は変更されることがありますのでご了承下さい。

図書目録進呈◆

新コロナシリーズ

(各巻B6判，欠番は品切です)

			頁	本体
2.	ギャンブルの数学	木下栄蔵著	174	1165円
3.	音戯話	山下充康著	122	1000円
4.	ケーブルの中の雷	速水敏幸著	180	1165円
5.	自然の中の電気と磁気	高木相著	172	1165円
6.	おもしろセンサ	國岡昭夫著	116	1000円
7.	コロナ現象	室岡義廣著	180	1165円
8.	コンピュータ犯罪のからくり	菅野文友著	144	1165円
9.	雷の科学	饗庭貢著	168	1200円
10.	切手で見るテレコミュニケーション史	山田康二著	166	1165円
11.	エントロピーの科学	細野敏夫著	188	1200円
12.	計測の進歩とハイテク	高田誠二著	162	1165円
13.	電波で巡る国ぐに	久保田博南著	134	1000円
14.	膜とは何か ―いろいろな膜のはたらき―	大矢晴彦著	140	1000円
15.	安全の目盛	平野敏右編	140	1165円
16.	やわらかな機械	木下源一郎著	186	1165円
17.	切手で見る輸血と献血	河瀬正晴著	170	1165円
19.	温度とは何か ―測定の基準と問題点―	櫻井弘久著	128	1000円
20.	世界を聴こう ―短波放送の楽しみ方―	赤林隆仁著	128	1000円
21.	宇宙からの交響楽 ―超高層プラズマ波動―	早川正士著	174	1165円
22.	やさしく語る放射線	菅野・関 共著	140	1165円
23.	おもしろ力学 ―ビー玉遊びから地球脱出まで―	橋本英文著	164	1200円
24.	絵に秘める暗号の科学	松井甲子雄著	138	1165円
25.	脳波と夢	石山陽事著	148	1165円
26.	情報化社会と映像	樋渡涓二著	152	1165円
27.	ヒューマンインタフェースと画像処理	鳥脇純一郎著	180	1165円
28.	叩いて超音波で見る ―非線形効果を利用した計測―	佐藤拓宋著	110	1000円
29.	香りをたずねて	廣瀬清一著	158	1200円
30.	新しい植物をつくる ―植物バイオテクノロジーの世界―	山川祥秀著	152	1165円
31.	磁石の世界	加藤哲男著	164	1200円

No.	書名	著者	頁	本体
32.	体を測る	木村雄治 著	134	1165円
33.	洗剤と洗浄の科学	中西茂子 著	208	1400円
34.	電気の不思議 ―エレクトロニクスへの招待―	仙石正和 編著	178	1200円
35.	試作への挑戦	石田正明 著	142	1165円
36.	地球環境科学 ―滅びゆくわれらの母体―	今木清康 著	186	1165円
37.	ニューエイジサイエンス入門 ―テレパシー,透視,予知などの超自然現象へのアプローチ―	窪田啓次郎 著	152	1165円
38.	科学技術の発展と人のこころ	中村孔治 著	172	1165円
39.	体を治す	木村雄治 著	158	1200円
40.	夢を追う技術者・技術士	CEネットワーク 編	170	1200円
41.	冬季雷の科学	道本光一郎 著	130	1000円
42.	ほんとに動くおもちゃの工作	加藤孜 著	156	1200円
43.	磁石と生き物 ―からだを磁石で診断・治療する―	保坂栄弘 著	160	1200円
44.	音の生態学 ―音と人間のかかわり―	岩宮眞一郎 著	156	1200円
45.	リサイクル社会とシンプルライフ	阿部絢子 著	160	1200円
46.	廃棄物とのつきあい方	鹿園直建 著	156	1200円
47.	電波の宇宙	前田耕一郎 著	160	1200円
48.	住まいと環境の照明デザイン	饗庭貢 著	174	1200円
49.	ネコと遺伝学	仁川純一 著	140	1200円
50.	心を癒す園芸療法	日本園芸療法士協会 編	170	1200円
52.	摩擦への挑戦 ―新幹線からハードディスクまで―	日本トライボロジー学会 編	176	1200円
53.	気象予報入門	道本光一郎 著	118	1000円
54.	続 もの作り不思議百科 ―ミリ,マイクロ,ナノの世界―	JSTP 編	160	1200円
55.	人のことば,機械のことば ―プロトコルとインタフェース―	石山文彦 著	118	1200円
56.	磁石のふしぎ	茂吉・早川 共著	112	1000円
57.	摩擦との闘い ―家電の中の厳しき世界―	日本トライボロジー学会 編	136	1200円
58.	製品開発の心と技 ―設計者をめざす若者へ―	安達瑛二 著	176	1200円
59.	先端医療を支える工学 ―生体医工学への誘い―	日本生体医工学会 編	168	1200円
60.	ハイテクと仮想の世界を生きぬくために	齋藤正男 著	144	1200円
61.	未来を拓く宇宙展開構造物 ―伸ばす,広げる,膨らませる―	角田博明 著	176	1200円
62.	科学技術の発展とエネルギーの利用	新宮原正三 著	154	1200円
63.	微生物パワーで環境汚染に挑戦する	椎葉究 著	144	1200円

定価は本体価格+税です。
定価は変更されることがありますのでご了承下さい。

図書目録進呈◆

技術英語・学術論文書き方，プレゼンテーション関連書籍

プレゼン基本の基本 －心理学者が提案するプレゼンリテラシー－
下野孝一・吉田竜彦 共著／A5／128頁／本体1,800円／並製

まちがいだらけの文書から卒業しよう 工学系卒論の書き方
－基本はここだ！－
別府俊幸・渡辺賢治 共著／A5／200頁／本体2,600円／並製

理工系の技術文書作成ガイド
白井　宏 著／A5／136頁／本体1,700円／並製

ネイティブスピーカーも納得する技術英語表現
福岡俊道・Matthew Rooks 共著／A5／240頁／本体3,100円／並製

科学英語の書き方とプレゼンテーション（増補）
日本機械学会 編／石田幸男 編著／A5／208頁／本体2,300円／並製

続 科学英語の書き方とプレゼンテーション
－スライド・スピーチ・メールの実際－
日本機械学会 編／石田幸男 編著／A5／176頁／本体2,200円／並製

マスターしておきたい 技術英語の基本 －決定版－
Richard Cowell・余　錦華 共著／A5／220頁／本体2,500円／並製

いざ国際舞台へ！ 理工系英語論文と口頭発表の実際
富山真知子・富山　健 共著／A5／176頁／本体2,200円／並製

科学技術英語論文の徹底添削 －ライティングレベルに対応した添削指導－
絹川麻理・塚本真也 共著／A5／200頁／本体2,400円／並製

技術レポート作成と発表の基礎技法（改訂版）
野中謙一郎・渡邉力夫・島野健仁郎・京相雅樹・白木尚人 共著
A5／166頁／本体2,000円／並製

知的な科学・技術文章の書き方 －実験リポート作成から学術論文構築まで－
中島利勝・塚本真也 共著
A5／244頁／本体1,900円／並製
日本工学教育協会賞（著作賞）受賞

知的な科学・技術文章の徹底演習
塚本真也 著
A5／206頁／本体1,800円／並製
工学教育賞（日本工学教育協会）受賞

定価は本体価格＋税です。
定価は変更されることがありますのでご了承下さい。

図書目録進呈◆